DNA Repair of Cancer Stem Cells

Lesley A. Mathews • Stephanie M. Cabarcas
Elaine M. Hurt
Editors

DNA Repair of Cancer Stem Cells

Springer

Editors
Dr. Lesley A. Mathews
Division of Pre-Clinical Innovation
National Center for Advancing Translational Sciences
National Institutes of Health
Rockville, Maryland
USA

Dr. Elaine M. Hurt
MedImmune
Oncology Research
Gaithersburg, Maryland
USA

Dr. Stephanie M. Cabarcas
Department of Biology
Gannon University
Erie, Pennsylvania
USA

ISBN 978-94-007-4589-6 ISBN 978-94-007-4590-2 (eBook)
DOI 10.1007/978-94-007-4590-2
Springer Dordrecht Heidelberg London New York

Library of Congress Control Number: 2012943246

Printed on acid-free paper

Springer is part of Springer Science+Business Media (www.springer.com)

Preface

In recent years, the strides made in understanding and elucidating both the origins and biological mechanisms responsible for driving cancer progression have been quite impressive. Specifically, the momentum that has coincided with the discovery and investigation of cancer stem cells (CSCs), tumor-initiating cells (TICs) or cancer-initiating cells (CICs) has been enormous. The investigation of every aspect of this deadly and lethal subpopulation has brought attention to its potential in a therapeutic light which we hope can translate into the clinic. The cancer stem cell hypothesis was first described with data from models of human leukemia by John E. Dick from the University of Toronto. The heterogeneity of human leukemia and the presence of stem cells in cancer was further translated into solid tumors by Al-Hajj et al. when they published a provocative paper in Proceedings of the National Academy of Sciences discussing the ability to distinguish tumorigenic (tumor-initiating) cancer cells from the nontumorigenic counterpart based on the expression of cell surface markers. The group reported that as little as 100 cells of this specific population were able to form a solid tumor when injected into the mammary fat pad of immunocompromised mice. The most critical aspect of this study was the data demonstrating that even tens of thousands of cells of the nontumorigenic cancer stem cell depleted fraction failed to produce a tumor.

Since this study, these cells have been heavily investigated and are now known to be the most aggressive cells within a solid tumor discovered to date. In recent years, many groups have demonstrated that in addition to being the most aggressive cells, they are highly resistant to current chemotherapy and radiation regimes employed in the clinic. The resistant nature of these cells has led many labs down the path of developing new therapies to eradicate them from patients. An interesting observation among our lab and others was that isolated CSCs express higher levels of DNA repair genes, and furthermore, lead to increased expression of crucial genes and pathways that contribute to their drug resistant characteristics. Thus, we have assembled a remarkable group of experts in both CSCs and DNA repair to discuss their research in light of the role of DNA repair genes and pathways in the CSC population. The common end goal is to contribute to the knowledge base and lead the field in investigating and studying additional mechanisms for potential therapies being designed to target this aggressive population of cells.

The concept of DNA repair conferring survival and progression is the overall theme of this book, and we believe provides a unique contribution to the CSC field in regards to developing new strategies to target this highly metastatic and resistant population. We hope this book can provide a foundation and support to future scientists and clinicians working in the field of cancer resistance and cancer stem cells.

Lesley Mathews
Stephanie M. Cabarcas

1. Al-Hajj M, Wicha MS, Benito-Hernandez A et al (2003) Prospective identification of tumorigenic breast cancer cells. Proc Natl Acad Sci USA 100(7):3983–3988
2. Mathews LA, Cabarcas SM, Hurt EM et al (2011) Increased expression of DNA repair genes in invasive human pancreatic cancer cells. Pancreas 40(5):730–739
3. Mathews LA, Cabarcas SM, Farrar WL (2011) DNA repair: the culprit for tumor-initiating cell survival? Cancer Metastasis Rev 30(2):185–197

Contents

Chapter 1
Introduction to Cancer Stem Cells

Chengzhuo Gao, Robert E. Hollingsworth and Elaine M. Hurt

Abstract A wealth of data points to the existence of a subset of tumor-initiating cells that have properties similar to stem cells, termed cancer stem cells (CSCs). CSCs are thought to be at the apex of a cellular hierarchy, where they are capable of differentiating into the other cells found within a tumor. They may also be responsible for both patient relapse due to their relative resistance to chemotherapy as well as metastasis. In recent years, much research has focused on these cells, their properties and potential targets within these cells for cancer treatment. This chapter will introduce the CSC theory, discuss important properties of these cells, and highlight the need to target them for improved patient outcome.

1.1 The Etiology of Cancer

Many hypotheses have been put forth through the years that attempt to explain the etiology of cancer. They have come from divergent fields of study; pathology, molecular biology and genetics but they all attempt to explain how a normal tissue can go from homeostatic equilibrium to something that grows without the checks and balances that govern normal biology. These theories include, but are not limited to, a viral basis of disease, clonal expansion, and the cancer stem cell hypothesis. While each of these theories attempts to explain the etiology of cancer, it is most likely that some of these independent theories work together to give rise to not only tumors, but tumors that are able to evade treatment strategies.

E. M. Hurt (✉)
MedImmune, LLC., Oncology Research MedImmune, Gaithersburg, MD 20878, USA
e-mail: hurte@medImmune.com

C. Gao · R. E. Hollingsworth
Oncology Research, MedImmune, LLC., Gaithersburg, MD 20878, USA

L. A. Mathews et al. (eds.), *DNA Repair of Cancer Stem Cells,*
DOI 10.1007/978-94-007-4590-2_1,

1.2 The Cancer Stem Cell Hypothesis

Since the early pathologists could look at tissues microscopically the origins and progression of cancer has been pondered. It was noted that advanced tumors often presented with diverse areas of differentiation, proliferation, invasion and vascularity. It was during this time of early microscopic inspection that led Cohnheim, a student of Virchow, to propose his embryonal rest hypothesis ([1] and references therein). Cohnheim had noted that cancer tissues, teratomas in particular, had many of the same properties as embryonic tissue. This lead him to speculate that carcinogenesis occurs from dormant remnants of the embryo that are later reactivated. However, this theory was largely ignored. The spirit of this hypothesis reappeard in the 1970s when Barry Pierce et al. examined differentiation in teratomas (reviewed in [1]) and determined that a stem cell was responsible for initiation of the teratomas. They then furthered these observations into a theory that all epithelial cancers arise as a result of differentiation-paused adult tissue stem cells [2].

On the heels of this hypothesis, came some of the first evidence that leukemias may have a CSC origin. John Dick and colleagues showed that a sub-population of acute myelogenous leukemia (AML) cells, which shared a phenotype with normal hematopoietic stem cells, could confer cancer when transplanted into immunocompromised mice. Furthermore, the cells that did not have the stem cell-phenotype could not transfer AML to recipient mice [3, 4]. Several years later, CSCs were identified in breast cancer [5], followed by a myriad of solid malignancies (discussed in more detail below in "1.4 Identification of CSCs").

1.3 Properties and Cell of Origin of CSCs

Cancer stem cells are so named because they share many of the same properties as normal stem cells. They are capable of tumorigenesis, self-renewal and can differentiate to form the heterogeneous cell types present in tumors (Fig. 1.1a). These functional properties led to the CSC moniker; however, many argue that "tumor-initiating cells" would be a better description of these cells. For the purposes of this book, we will refer to these cells as CSCs, where a CSC has been defined as "a cell within a tumor that possess the capacity to self-renew and to cause the heterogeneous lineages of cancer cells that comprise the tumor" [6]. It has also been observed that CSCs are relatively resistant to chemotherapy and therefore may be responsible for patient relapse following treatment (Fig. 1.1b, discussed in detail in Chap. 3). Thus, the CSC hypothesis attempts to explain several observations of tumors, including the frequency at which tumor cells can give rise to new tumors, the generation of cells with multiple genetic alterations, and the heterogeneity of cell types present within tumors.

In theory, cancer arises from a single cell that has somehow subverted normal growth restrictions. However, experimental evidence has shown a requirement of many cells in order to seed a tumor. For example the growth of tumor cells in immunocompromised mice, typically requires that 1–10 million cells are implanted in

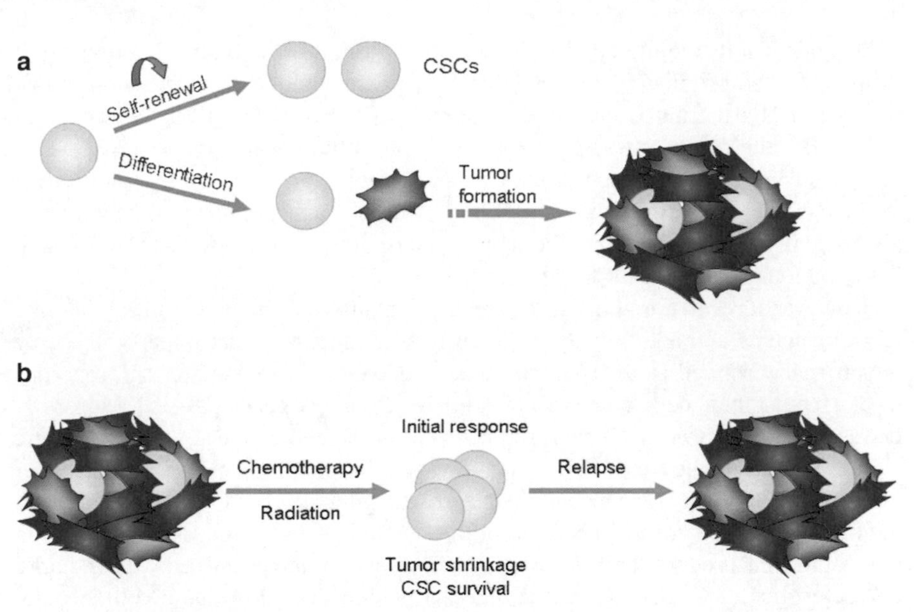

Fig. 1.1 Schematic representation of CSC-driven tumor formation. (**a**) CSCs can either divide symmetrically (self-renewal) to give rise to two CSCs or they can divide asymmetrically (differentiation) giving rise to one CSC and one differentiated progeny. The ability to give rise to a tumor, to self-renewal and differentiate into the heterogeneous cell populations found in the tumor are the defining characteristic of a CSC. The differentiated progeny of a CSC are often more proliferative than the CSC itself, but they have a finite replicative capacity. (**b**) CSCs are resistant to conventional therapies, including chemotherapy and radiation. These treatments can eliminate the more rapidly dividing differentiated cells but leave behind the CSC (discussed in more detail in Chap. 3). Once treatment is stopped, the remaining CSCs can begin to divide again and form a new tumor resulting in patient relapse

order to see tumor formation. This observation has many possible explanations, including injury to cells upon injection, the requirement for the right microenvironment (and the difference of this between human and mouse), as well as the requirement to have a variety of cell types present in order to efficiently induce tumor formation. However, even in experiments where human cancer patients were reinjected with their own tumor cells at different sites within their bodies, it took large numbers of cells in order for a new tumor to establish. Again, this could be consistent with the role of the tumor microenvironment, this time from one site to another; it is also consistent with the notion that only a small proportion of cells are capable of giving rise to a tumor. The idea that only a small proportion of the cells found within a tumor are capable of giving rise to a tumor is consistent with the CSC hypothesis. The frequency of CSCs in most cases has been reported to be low (typically less than 5 %) except in the case of melanoma where a variety of cells were shown to have equally high tumorigenicity [7, 8].

The origin of the cancer stem cell is still under debate. The requirement for multiple genetic insults in order to drive tumor formation has been recognized for a long time [9]. It is suggested that in order for a cell to live long enough to sustain the genetic insults required to drive tumor formation, it is likely that a normal adult stem cell is the originator of a CSC phenotype. Mathematical models support this theory [10]. Furthermore, the leukemic stem cells (LSCs) of CML patients express BCR-ABL, the common translocation that drives cellular transformation [11]. This provides experimental evidence that CSCs sustain the genetic insults that are seen to drive carcinogenesis. The identification in CSCs of oncogenic driver mutations is also beginning to emerge for solid tumors. For example, the *TMPRSS:ERG* fusion has been found in prostate CSCs [12].

However, it is still possible that a more differentiated cell undergoes these genomic rearrangements and additionally picks up further mutations that impart the ability to self-renew as well as differentiate. Indeed fibroblasts can acquire the properties of pluripotent stem cells with the activation of just a few genes [13–15]. Moreover, it was recently shown that a catastrophic event in the cell can lead to many genetic alterations at a single time [16]. It may also be that environmental cues can trigger CSC properties even independent of DNA rearrangements. For example, it has been noted that cells undergoing epithelial-to-mesenchymal transition (EMT), a normal developmental process that promotes cancer invasion and metastasis, can acquire characteristics of CSCs [17]. These ilines of evidence suggest that a cell does not need to be long-lived in order to sustain many genetic alterations and that CSC properties can be bestowed by biological processes, and would therefore argue that any cell may be the fodder for CSCs.

1.4 Isolation of CSCs

Currently, there are several commonly used approaches for the isolation of cancer stem cells, including: (1) sorting of a side population (SP) by flow cytometry based on Hoechst dye efflux, (2) sorting of CSCs by flow cytometry based on cell surface marker expression, (3) enriching of CSCs by non-adherent sphere culture and (4) sorting of CSCs by flow cytometry based on aldehyde dehydrogenase (ALDH) activity (Fig. 1.2). All of these approaches enrich for CSCs to varying degrees, and each of them has its own advantages and limitations, which will be discussed below.

1.4.1 Side Populations

Goodell and colleagues, while analyzing murine bone marrow cells, discovered a small and distinct subset of whole bone marrow cells that were unstained by Hoechst 33342, a vital dye [18]. This Hoechst 33342 low population is termed SP. They found that the SP had the phenotypic markers of multipotential hematopoietic stem cells and were able to repopulate the bone marrow. Following their work, the SP has been extended to a variety of cancer types, including leukemia [19, 20], ovarian cancer [21], hepatocellular carcinoma [22], brain tumors [23–25], lung cancer [26, 27], thyroid cancer [28], nasopharyngeal carcinoma [29], mesenchymal tumors

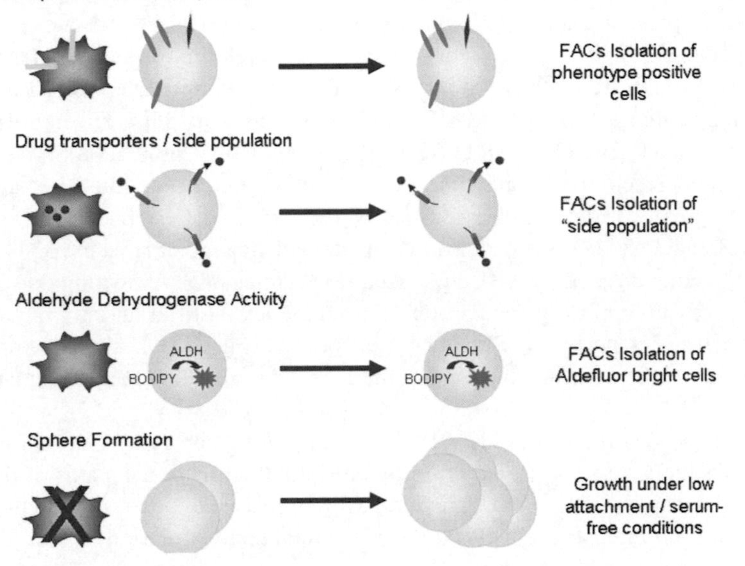

Fig. 1.2 Methods of CSC isolation. CSCs have been isolated using one of four methods. This is a pictorial representation of these isolation techniques and how they are used to identify the CSC versus the non-CSCs

[30], colon cancer [31], prostate cancer [32], breast cancer [33–35], head and neck cancer [36] and other cancers [37–40]. The SP cells purified from these tumor types harbor cancer stem cell-like cells with properties such as a "stemness" gene signature [23, 26, 28, 29, 35], self-renewal capacity [25, 26, 28, 29, 33, 35] and tumorigenicity [19, 21, 24–26, 28, 29, 35].

The ATP dependent transporter, ATP-binding cassette sub-family G member 2 (ABCG2, BRCP1), is generally believed to be responsible for Hoechst 33342 efflux by the SP. Several lines of evidence support this hypothesis. First, ABCG2 knockout mice show significantly decreased numbers of SP cells in both the bone marrow as well as in skeletal muscle [41]. Second, overexpression of ABCG2 dramatically increases the SP percentage of bone-marrow cells and reduces maturing progeny both *in vitro* and *in vivo* [42]. Third, ABCG2 is highly expressed in a wide variety of stem cells including the SP cells of neuroblastoma patients [23, 42]. Fourth, ABCG2 may be responsible for conferring drug resistance to the SP and CSCs. ABCG2 is a multidrug resistant pump expressed at variable levels in cancer cells, which can bind and expel cytotoxic drugs [43]. Thus, ABCG2 may lower intracellular levels of anticancer agents below the threshold for cell death in tumors, leaving resistant cells to repopulate the tumor. Indeed, inhibition of the ABC transporters sensitized SP cells of various cancer types to chemotherapeutic agents [44, 45].

However, there is not always a correlation between ABCG2 expression and the SP phenotype. For example, erythroblasts highly express the ABC transporter ABCG2 but do not have an SP phenotype [46]. Furthermore, the expression of ABCG2, and the presence of the SP in general, do not always define the CSC population. For instance, in prostate cancer both purified ABCG2^{+} and ABCG2^{-} cancer cells have similar tumorigenicity to the SP cells *in vivo*, although the SP cells express higher level of ABCG2 mRNA than the non-SP cells [32]. Additionally, ABCG2 is not the only multi-drug resistance gene identified in SP stem-like cells. The expression of P-glycoprotein (ABCB1, MDR1) is also significantly up-regulated in SP cells of an oral squamous cell carcinoma cell line [47], although Feuring-Buske et al. found that there is no correlation between the expression of ABCB1 and the SP in acute myeloid leukemia [19]. Instead, Zhou et al. demonstrated that Hoechst 33342 efflux activity is compensated by ABCG2 in Abcb1 null mice [42], indicating that SP cells may utilize different drug transportation machineries in different environments. In addition, the presence of the SP may instead be the result of inefficient dye uptake as a reflection of the presence of largely quiescent cells, another characteristic of stem cells [26, 48].

Compared to other methods, isolation of the side population has two advantages. First, it carries additional information about the functional status of the cells, since this assay is based on an active metabolic process. Second, it is highly sensitive, with even rare SP events (<0.5 % of the total cell population) detected within heterogeneous samples [49].

But this method also has many disadvantages. Some of the limitations of this method have to do with the Hoechst 33342 dye itself. Owing to the fact that Hoechst 33342 is a DNA binding dye, it is toxic to cells. Shen and colleagues found that Hoechst 33342 staining for a prolonged periods of time increases apoptosis in C6 cells [25]. Furthermore, it is highly sensitive to slight variations in staining conditions. Hoechst concentration, the staining time, and the staining temperatures all are critical for the success of this approach.

Other disadvantages of SP isolation are due to the ability of this method to accurately define and purify CSCs. Importantly, the SP is not always necessary or sufficient for a CSC phenotype. In glioblastoma multiforme (GBM), the SP from the GBM lines did not enrich for stem-like activity *in vitro*, and tumorigenicity was lower in sorted SP compared with non-SP and parental cells [50]. Equally important, is that SP cells represent a heterogeneous cell population. Wan et al. demonstrated that SP cells purified from a laryngeal cancer cell line does harbor cancer stem cell-like properties, but they are heterogeneous indicating that SP cells are not identical to stem cells [51]. Combining SP detection with cell surface marker selection, may lead to a more efficient and reliable isolation of CSCs.

Despite the limitations existing in SP isolation, the presence of the SP population has some clinical relevance in certain disease indications. For example, the SP population can be identified in gastric cancer tissue and correlates with patient survival [37]. A limited clinical study in ovarian cancer also revealed a higher SP frequency in recurrent or metastatic tumors compared with primary tumors, suggesting a good correlation between the presence of SP and recurrence in ovarian cancer [52].

Besides the use in prognosis of cancer, the SP may also serve as a potential therapeutic target for cancer. Several studies have pioneered the possibility of specifically targeting SP cells by exploiting pathways involved in drug resistance and differentiation [53]. For instance, Praveen et al. have isolated SP cells from multiple cancer cell lines and found that these SP cells are resistant to cytochrome C release and apoptosis. Based on this finding, they developed a high-throughput imaging assay, in which the cytochrome C-EGFP translocation is monitored in the sorted SP cells. Through this assay, the heat shock protein 90 inhibitors have been identified to sensitize the SP cells to some antitumor agents, such as cisplatin [54]. Moreover, an autologous vaccine for B-cell chronic lymphocytic leukemia (B-CLL) was made using a patient's SP cells. Following vaccination, the study showed an increase in (B-CLL)-reactive T-cells followed by a corresponding decline in circulating B-CLL SP cells [55]. This indicates that the SP may be a valid ground for cancer therapy.

1.4.2 Surface Markers

Initially used to identify and isolate normal stem cells, surface markers are now extensively used for the identification and isolation of CSCs in many malignances. Lapidot and colleagues were the first to isolate leukemia-initiating cells based on cell-surface marker expression and found $CD34^+$ $CD38^-$ cells, but not the $CD34^+$ $CD38^+$ and $CD34^-$ cells, harbored serial leukemic transplantation potential [3]. Following this initial prospective isolation of leukemia stem cells, breast CSCs were identified as $CD44^+$ $CD24^{-/low}$ by Al-Hajj and colleagues [5]. Later, based on their surface marker expression, CSCs have been isolated from various tumors (Table 1.1). Among all the surface markers used for CSCs isolation, CD133 and CD44 are the most commonly used in a variety of tumor types.

CD133 (Prominin 1), a five transmembrane glycoprotein, was originally identified both in the neuroepithelium and in various other epithelia of the mouse embryo [77]. Later, a novel monoclonal antibody recognizing the AC133 antigen, a glycosylation-dependent epitope of CD133, detected that CD133 is restricted in $CD34^+$ progenitor populations from adult blood, bone marrow and fetal liver cells [78]. In addition, CD133 expression is rapidly down regulated upon cell differentiation [79]. These characteristics of CD133 make it a unique cell surface marker for the identification and isolation of various CSCs (Table 1.1). The biological function of CD133 is still largely unknown. A single nucleotide deletion, which caused the truncation of CD133, is linked to an inherited form of human retinal degeneration [80]. A recent report has linked CD133 with endocytosis. In this study, CD133 knockdown improved Alexa488-transferrin (Tf) uptake in Caco-2 cells, while cell treatment with the AC133 antibody resulted in down regulated Tf uptake, [81]. Despite its utility as a marker of CSCs, CD133 does not appear to play a significant role in the maintenance of at least some CSCs. In colon cancer cells isolated from patients, CD133 knockdown did not affect their tumorigenicity *in vitro* and *in vivo* [82]. Instead, CD44 knockdown prevented tumor formation of the same cells.

CD44, a single transmembrane glycoprotein, is a major component of the extracellular matrix [83]. Besides acting as an adhesion molecule, it also functions as

Table 1.1 Cell surface markers used in the isolation of various CSCs

Tumor type	Phenotype	Reference
AML	$CD34^+CD38^-$	[3, 4]
Breast	$CD44^+CD24^-$	[5]
Brain	$CD133^+$	[56, 57]
	SSEA-1	[58]
Prostate	$CD44^+CD133^+\alpha_2\beta_1^{hi}$	[59]
	$CD44^+CD24^-$	[60]
Head and neck	$CD44^+$	[61]
Liver	$CD133^+$	[62]
	$CD90^+$	[63]
Colon	$CD133^+$	[64, 65]
	$EpCAM^{hi}CD44^+$	[66]
	$CD44^+/CD166^+$	[67]
Pancreatic	$CD44^+CD24^-EpCAM^+$	[68]
	$CD133^+$	[69]
Squamous cell carcinoma	Podoplanin	[70]
Lung	$CD133^+$	[71]
Melanoma	$ABCB5^+$	[72]
Gastric	$CD44^+$	[73]
Ovarian	$CD133^+$	[74]
	$CD44^+/CD177^+$	[75]
	$CD44^+$	[76]

a principle receptor for hyaluronan (HA) [84]. HA is enriched in the pericellular matrices of many malignant human tumors and plays an important role in tumor progression via regulation of receptor tyrosine kinases (RTKs), such as ERBB2 and EGFR [85]. Thus, as a critical receptor for HA, CD44 plays an important role in cell proliferation and survival via activation of the MAPK and PI3K/AKT pathways, respectively [85]. Furthermore, CD44 also plays an important role in the invasion of a variety of cancer cells, including breast [86], prostate [87], hepatoma [88], and mesotheliomas [89], and has been significantly correlated with the circulating prostate tumor cells [90]. Therefore, CD44 stands out as a surface marker for CSCs, as first shown by Al-hajj and colleagues in breast cancer [5]. Following their work, CD44 has been utilized as a surface marker to isolate CSCs from a variety of different tumors (Table 1.1).

CD44 was also explored as a potential diagnostic target for cancer detection as well as a drug target for cancer therapy [91–93]. For instance, in 2003, the humanized anti-CD44 antibody (bivatuzumab) labeled with rhenium-186 was used in phase I studies in patients with head and neck squamous cell carcinoma (HNSCC) [94, 95]. In this trial, these radiolabeled CD44 antibodies showed promising anti-tumor effects with low toxicity. Further, a different trial with the non-radiolabeled CD44 antibody (bivatuzumab mertansine) also had good patient response rates, although the development of this drug was terminated due to the death of a patient [96]. Recently, CD44 was also shown to target CSCs. In 2006, Jin et al. found that interruption of CSCs interaction with their microenvironment by monoclonal antibody directed against CD44 markedly decreased the number of the AML LSCs *in vivo*, indicating a key regulatory role of CD44 in AML LSCs [97].

While CD44 and CD133 appear to have broad tumor applicability, the choice of the cell surface markers tends to be tissue specific and is often based on previous knowledge of the development of that tissue. For instance, CD34 and CD38, the markers used for isolation of AML CSCs, are also the markers used to identify normal early hematopoietic progenitor cells [3, 4]. Another good example is CD138. CD138 is a marker for terminally-differentiated B cells (plasma cells). It has been shown that CSCs from multiple myeloma, a plasma cell malignancy, are $CD138^-$.

Compared to other methods of isolation, cell surface marker isolation has a major advantage of obtaining a precise population. However, the selection of which surface markers to use is one of the greatest pitfalls of this approach. As discussed earlier, many times markers selected are based on previous knowledge of the development of the tissue. For this reason, it may not be easy to find the right markers. For example, until recently there were no markers identified for CSCs of human primary gastric tumors [98]. Furthermore, the choice of markers made by researchers has not been unified, even within the same tumor type. As shown in Table 1.1, different markers have been used for the same tumor type. To further complicate the picture, many times there have not been careful comparisons done within the same study of all the proposed CSC markers in order to definitively test which marker combination is the best at identifying CSCs. A further complication is that the surface markers may be heterogeneous between patients even within the same tumor type. A recent study from 16 AML patients shows that the majority of LSCs are in the minor $CD34^+$ $CD38^-$ fraction in 50 % of the subjects, and in the $CD34^+$ $CD38^+$ fraction in the other 50 % [99]. Similar findings were also obtained from breast cancer patients. When Park et al. [92] evaluated the expression of stem cell-related markers at the cellular level in human breast tumors of different subtypes and histologic stages, they found that the cancer stem cell markers vary according to tumor subtype and histologic stage [100]. Ali et al. also demonstrated that breast CSC markers, such as CD44/CD24, ALDH and ITGA6, do not identify identical subpopulations in primary tumors [101].

Other limitations of choosing surface markers for the identification of CSCs are methodological. First, a large number of cells is required to sort and the number of CSCs identified by this approach is usually low ($<1 - 10$ %). Furthermore, when using tumor samples, the cells must first be dissociated typically with collagenase and/or other proteolytic enzymes. This dissociation step may damage the presentation of the cell surface antigens [102, 103].

1.4.3 Nonadherent Sphere Culture

Initially, nonadherent sphere culture was used to culture neural stem cells [104]. The cells isolated from the stratum of adult mouse by this method could generate both neurons and astrocytes. Later, it was shown that purified stem cells were capable of growing as spheres. For instance, $CD133^+$ cells isolated from normal human fetal brain formed spheres *in vitro* [105]. These observations were then extended to cancerous tissues where further studies demonstrated that $CD133^+$ cells from the human brain tumors are also capable of forming neurospheres [57]. Since then, the ability of CSCs to form spheres in culture has been shown for most solid malignancies, including breast [106], melanoma [107], pancreatic [108], prostate [109], ovarian [75] and colon CSCs [110].

Most importantly, researchers found that the nonadherent sphere culture condition can enrich cells with CSC phenotypes [106]. They demonstrated that, in breast cancer, the CSC population with $CD44^+$ $CD24^-$ phenotype increased more than two-fold after culturing under nonadherent sphere culture conditions. In addition, the spheres were more tumorigenic. Since then, researchers have used sphere to formation to enrich CSCs from brain [56, 57, 111, 112], colon [110], pancreas [108], bone sarcomas [113] melanomas [107] and prostate [109]. The enrichment of CSCs by growing cells under sphere culture conditions has been confirmed by surface marker expression in all mentioned cases, except bone sarcoma, in which the surface marker remain to be determined.

Compared to the other methods of isolating CSCs, this method is easy to perform and allows researchers to obtain a larger number of CSCs. However, like all methods of isolation, this method also suffers from several disadvantages. The spheres are a heterogeneous population of cells containing both CSCs and non-CSCs. For example it has been shown that only a portion of the spheres are capable of self-renewal [106, 114]. Immunostaining of spheres from prostate cancer cell lines indicated that the spheres are heterogeneous for CSC markers [115]. Also, the conditions of sphere formation are critical to the overall success of CSC enrichment. In neurosphere cultures, it has been shown that the composition of spheres can be different due to the differences in sphere size, passage, culture medium, and technique [116].

1.4.4 ALDH Activity

In addition to the above three isolation methods, isolation of CSCs based on their aldehyde dehydrogenase (ALDH) activity is also commonly used. ALDH is a detoxifying enzyme responsible for the oxidation of intracellular aldehydes [117]. It functions in drug resistance, cell differentiation and oxidative stress response [118]. ALDH activity may be easily assessed in living cells using the ALDEFLUOR kit (Stem Cell Technologies). This kit utilizes an ALDH substrate BODIPY aminoacetaldehyde (BAAA), which is converted in the cytoplasm into a florescent molecule by ALDH enzymatic activity [119]. Recent studies have demonstrated a positive correlation between ALDH expression and overall survival of patients with different cancers [120]. To date, ALDH activity has been used to identify and isolate CSCs from AML [121], breast [122], melanoma [123], prostate [124], liver [71], ovarian [125], lung [126] and osteosarcoma [127].

Like other isolation methods, ALDH is also not a universal marker for cancer stem cells in any tumor type. In a CSC marker profiling study, the breast cancer cells selected by ALDH activity did not always identify the most tumorigenic cells [128]. Furthermore, in prostate cancer, Yu et al. [14] found that $ALDH^{lo}$ $CD44^{-}$ cells were also able to develop tumors with similar frequency as the $ALDH^{hi}CD44^{+}$ cells, although with longer latency periods [129].

To date, there is not a universal isolation method in the CSC field. Researchers have to select their isolation method depending on the purpose of their studies, and sample source. Recently, Zhou et al. demonstrated that SP rather than CD133(+) cells indicate enriched tumorigenicity in hTERT-immortalized primary prostate cancer cells [130]. Sometimes, two different methods may need to be combined to isolate CSCs. For instance, combining SP determination with cell-surface marker phenotyping leads to efficient, reliable characterization of the HSC subset [131]. Further, Eppert and colleagues found that AML LSCs from different patient samples have diverse surface marker profiles and frequency [99]. However, these LSCs shared a core gene signature, indicating that the stemness gene signature might be a more accurate way to identify CSCs.

1.5 Implications of the CSC Hypothesis for Cancer Treatment

The discovery of cancer stem cells has important implications for oncology, and many groups are now working to understand and exploit CSC biology to improve cancer treatment. Despite tremendous progress in the identification of new cancer targets and drug development, most cancers relapse after treatment and the goal of increasing overall survival of cancer patients has been largely unmet by most new drugs [132]. One explanation for this lack of sustained effect is that most current therapies do not effectively inhibit CSCs.

Indeed, CSCs are resistant to chemo- and radiation therapies that effectively kill other cells that comprise the tumor (discussed in detail in Chap. 3). It is for this reason that CSCs are thought to regenerate cancer when therapy is discontinued. This hypothesis holds that, because CSCs are a small population within the tumor mass, drugs developed to kill the cells comprising the bulk of the tumor are initially effective in shrinking tumor size but do not eradicate the cellular source and tumors with higher CSC content and perhaps harboring additional drug-resistance mutations regrow. This has been termed the "dandelion hypothesis," analogous to the regrowth of a weed if the root is not destroyed [133]. Although this hypothesis has yet to be fully confirmed using a CSC-directed therapy in cancer patients, numerous studies have correlated CSCs with poor prognosis in both leukemia and solid malignancies. For example, glioblastoma multiforme patients whose tumors bear a relatively high proportion of $CD133^{+}$ cells, which have been found to possess characteristics of CSCs, suffer from decreased response to therapy, higher malignancy, and significantly lower survival time [134]. Additionally, gene signatures derived from CSCs are themselves independent predictors of patient survival [99, 135] indicating the relevance of targeting these cells.

1.6 Concluding Remarks

A consensus is emerging that the CSC population represents a distinct and clinically important population. However, there is still a further need to determine the precise definition and cellular source of CSCs, to elucidate the appropriate cell markers to isolate them, and to understand the pathways contributing their biological behavior. In the end there is growing hope that this understanding will translate into clinically useful treatments against cancer.

References

1. Pierce GB, Johnson LD (1971) Differentiation and cancer. In Vitro 7:140–145
2. Sell S, Pierce GB (1994) Maturation arrest of stem cell differentiation is a common pathway for the cellular origin of teratocarcinomas and epithelial cancers. Lab Invest 70:6–22
3. Lapidot T, Sirard C, Vormoor J et al (1994) A cell initiating human acute myeloid leukaemia after transplantation into SCID mice. Nature 367:645–648
4. Bonnet D, Dick JE (1997) Human acute myeloid leukemia is organized as a hierarchy that originates from a primitive hematopoietic cell. Nat Med 3:730–737
5. Al-Hajj M, Wicha MS, Benito-Hernandez A, Morrison SJ, Clarke MF (2003) Prospective identification of tumorigenic breast cancer cells. Proc Natl Acad Sci USA 100:3983–3988
6. Clarke MF, Dick JE, Dirks PB et al (2006) Cancer stem cells—perspectives on current status and future directions: AACR Workshop on cancer stem cells. Cancer Res 66:9339–9344
7. Ishizawa K, Rasheed ZA, Karisch R et al (2010) Tumor-initiating cells are rare in many human tumors. Cell Stem Cell 7:279–282
8. Shackleton MJ, Quintana E, Fullen DR, Sabel MS, Johnson TM (2009) Melanoma: do we need a hatchet or a scalpel? Arch Dermatol 145:307–308
9. Nordling CO (1953) A new theory on cancer-inducing mechanism. Br J Cancer 7:68–72
10. Ashkenazi R, Gentry SN, Jackson TL (2008) Pathways to tumorigenesis—modeling mutation acquisition in stem cells and their progeny. Neoplasia 10:1170–1182
11. Maguer-Satta V, Petzer AL, Eaves AC, Eaves CJ (1996) BCR-ABL expression in different subpopulations of functionally characterized Ph+ $CD34^{+}$ cells from patients with chronic myeloid leukemia. Blood 88:1796–1804
12. Birnie R, Bryce SD, Roome C et al (2008) Gene expression profiling of human prostate cancer stem cells reveals a pro-inflammatory phenotype and the importance of extracellular matrix interactions. Genome Biol 9:R83
13. Takahashi K, Yamanaka S (2006) Induction of pluripotent stem cells from mouse embryonic and adult fibroblast cultures by defined factors. Cell 126:663–676
14. Yu J, Vodyanik MA, Smuga-Otto K et al (2007) Induced pluripotent stem cell lines derived from human somatic cells. Science 318:1917–1920
15. Takahashi K, Tanabe K, Ohnuki M et al (2007) Induction of pluripotent stem cells from adult human fibroblasts by defined factors. Cell 131:861–872
16. Stephens PJ, Greenman CD, Fu B et al (2011) Massive genomic rearrangement acquired in a single catastrophic event during cancer development. Cell 144:27–40
17. Mani SA, Guo W, Liao MJ et al (2008) The epithelial-mesenchymal transition generates cells with properties of stem cells. Cell 133:704–715
18. Goodell MA, Brose K, Paradis G, Conner AS, Mulligan RC (1996) Isolation and functional properties of murine hematopoietic stem cells that are replicating in vivo. J Exp Med 183:1797–1806

19. Feuring-Buske M, Hogge DE (2001) Hoechst 33342 efflux identifies a subpopulation of cytogenetically normal CD34(+)CD38(−) progenitor cells from patients with acute myeloid leukemia. Blood 97:3882–3889
20. Wulf GG, Wang RY, Kuehnle I et al (2001) A leukemic stem cell with intrinsic drug efflux capacity in acute myeloid leukemia. Blood 98:1166–1173
21. Szotek PP, Pieretti-Vanmarcke R, Masiakos PT et al (2006) Ovarian cancer side population defines cells with stem cell-like characteristics and Mullerian Inhibiting Substance responsiveness. Proc Natl Acad Sci USA 103:11154–11159
22. Chiba T, Kita K, Zheng YW et al (2006) Side population purified from hepatocellular carcinoma cells harbors cancer stem cell-like properties. Hepatology 44:240–251
23. Hirschmann-Jax C, Foster AE, Wulf GG et al (2004) A distinct "side population" of cells with high drug efflux capacity in human tumor cells. Proc Natl Acad Sci USA 101:14228–14233
24. Setoguchi T, Taga T, Kondo T (2004) Cancer stem cells persist in many cancer cell lines. Cell Cycle 3:414–415
25. Shen G, Shen F, Shi Z et al (2008) Identification of cancer stem-like cells in the C6 glioma cell line and the limitation of current identification methods. In Vitro Cell Dev Biol Anim 44:280–289
26. Ho MM, Ng AV, Lam S, Hung JY (2007) Side population in human lung cancer cell lines and tumors is enriched with stem-like cancer cells. Cancer Res 67:4827–4833
27. Salcido CD, Larochelle A, Taylor BJ, Dunbar CE, Varticovski L (2010) Molecular characterisation of side population cells with cancer stem cell-like characteristics in small-cell lung cancer. Br J Cancer 102:1636–1644
28. Mitsutake N, Iwao A, Nagai K et al (2007) Characterization of side population in thyroid cancer cell lines: cancer stem-like cells are enriched partly but not exclusively. Endocrinology 148:1797–1803
29. Wang J, Guo LP, Chen LZ, Zeng YX, Lu SH (2007) Identification of cancer stem cell-like side population cells in human nasopharyngeal carcinoma cell line. Cancer Res 67:3716–3724
30. Wu C, Wei Q, Utomo V et al (2007) Side population cells isolated from mesenchymal neoplasms have tumor initiating potential. Cancer Res 67:8216–8222
31. Sussman RT, Ricci MS, Hart LS, Sun SY, El-Deiry WS (2007) Chemotherapy-resistant side-population of colon cancer cells has a higher sensitivity to TRAIL than the non-SP, a higher expression of c-Myc and TRAIL-receptor DR4. Cancer Biol Ther 6:1490–1495
32. Patrawala L, Calhoun T, Schneider-Broussard R, Zhou J, Claypool K, Tang DG (2005) Side population is enriched in tumorigenic, stem-like cancer cells, whereas ABCG2^{+} and ABCG2^{-} cancer cells are similarly tumorigenic. Cancer Res 65:6207–6219
33. Christgen M, Ballmaier M, Bruchhardt H, von Wasielewski R, Kreipe H, Lehmann U (2007) Identification of a distinct side population of cancer cells in the Cal-51 human breast carcinoma cell line. Mol Cell Biochem 306:201–212
34. Nakanishi T, Chumsri S, Khakpour N et al (2010) Side-population cells in luminal-type breast cancer have tumour-initiating cell properties, and are regulated by HER2 expression and signalling. Br J Cancer 102:815–826
35. Engelmann K, Shen H, Finn OJ (2008) MCF7 side population cells with characteristics of cancer stem/progenitor cells express the tumor antigen MUC1. Cancer Res 68:2419–2426
36. Tabor MH, Clay MR, Owen JH et al (2011) Head and neck cancer stem cells: the side population. Laryngoscope 121:527–533
37. Schmuck R, Warneke V, Behrens HM, Simon E, Weichert W, Rocken C (2011) Genotypic and phenotypic characterization of side population of gastric cancer cell lines. Am J Pathol 178:1792–1804
38. Yanamoto S, Kawasaki G, Yamada S et al (2011) Isolation and characterization of cancer stem-like side population cells in human oral cancer cells. Oral Oncol 47:855–860
39. Yang M, Yan M, Zhang R, Li J, Luo Z (2011) Side population cells isolated from human osteosarcoma are enriched with tumor-initiating cells. Cancer Sci 102:1774–1781
40. Zhang SN, Huang FT, Huang YJ, Zhong W, Yu Z (2011) Characterization of a cancer stem cell-like side population derived from human pancreatic adenocarcinoma cells. Tumori 96:985–992

41. Zhou S, Morris JJ, Barnes Y, Lan L, Schuetz JD, Sorrentino BP (2002) Bcrp1 gene expression is required for normal numbers of side population stem cells in mice, and confers relative protection to mitoxantrone in hematopoietic cells in vivo. Proc Natl Acad Sci USA 99:12339–12344
42. Zhou S, Schuetz JD, Bunting KD et al (2001) The ABC transporter Bcrp1/ABCG2 is expressed in a wide variety of stem cells and is a molecular determinant of the side-population phenotype. Nat Med 7:1028–1034
43. Mao Q, Unadkat JD (2005) Role of the breast cancer resistance protein (ABCG2) in drug transport. AAPS J 7:E118–133
44. Loebinger MR, Giangreco A, Groot KR et al (2008) Squamous cell cancers contain a side population of stem-like cells that are made chemosensitive by ABC transporter blockade. Br J Cancer 98:380–387
45. Katayama R, Koike S, Sato S, Sugimoto Y, Tsuruo T, Fujita N (2009) Dofequidar fumarate sensitizes cancer stem-like side population cells to chemotherapeutic drugs by inhibiting ABCG2/BCRP-mediated drug export. Cancer Sci 100:2060–2068
46. Yamamoto K, Suzu S, Yoshidomi Y, Hiyoshi M, Harada H, Okada S (2007) Erythroblasts highly express the ABC transporter Bcrp1/ABCG2 but do not show the side population (SP) phenotype. Immunol Lett 114:52–58
47. Yajima T, Ochiai H, Uchiyama T, Takano N, Shibahara T, Azuma T (2009) Resistance to cytotoxic chemotherapy-induced apoptosis in side population cells of human oral squamous cell carcinoma cell line Ho-1-N-1. Int J Oncol 35:273–80
48. Bhatt RI, Brown MD, Hart CA et al (2003) Novel method for the isolation and characterisation of the putative prostatic stem cell. Cytometry A 54:89–99
49. Golebiewska A, Brons NH, Bjerkvig R, Niclou SP (2011) Critical appraisal of the side population assay in stem cell and cancer stem cell research. Cell Stem Cell 8:136–147
50. Broadley KW, Hunn MK, Farrand KJ et al (2011) Side population is not necessary or sufficient for a cancer stem cell phenotype in glioblastoma multiforme. Stem Cells 29:452–461
51. Wan G, Zhou L, Xie M, Chen H, Tian J (2010) Characterization of side population cells from laryngeal cancer cell lines. Head Neck 32:1302–1309
52. Hosonuma S, Kobayashi Y, Kojo S et al (2011) Clinical significance of side population in ovarian cancer cells. Hum Cell 24:9–12
53. Moserle L, Ghisi M, Amadori A, Indraccolo S (2009) Side population and cancer stem cells: therapeutic implications. Cancer Lett 288:1–9
54. Sobhan PK, Seervi M, Joseph J et al (2011) Identification of heat shock protein 90 inhibitors to sensitize drug resistant side population tumor cells using a cell based assay platform. Cancer Lett 317:78–88
55. Foster AE, Okur FV, Biagi E et al (2010) Selective elimination of a chemoresistant side population of B-CLL cells by cytotoxic T lymphocytes in subjects receiving an autologous hCD40L/IL-2 tumor vaccine. Leukemia 24:563–572
56. Singh SK, Clarke ID, Terasaki M et al (2003) Identification of a cancer stem cell in human brain tumors. Cancer Res 63:5821–5828
57. Singh SK, Hawkins C, Clarke ID et al (2004) Identification of human brain tumour initiating cells. Nature 432:396–401
58. Son MJ, Woolard K, Nam DH, Lee J, Fine HA (2009) SSEA-1 is an enrichment marker for tumor-initiating cells in human glioblastoma. Cell Stem Cell 4:440–452
59. Collins AT, Berry PA, Hyde C, Stower MJ, Maitland NJ (2005) Prospective identification of tumorigenic prostate cancer stem cells. Cancer Res 65:10946–10951
60. Hurt EM, Kawasaki BT, Klarmann GJ, Thomas SB, Farrar WL (2008) $CD44^{+}$ CD24(−) prostate cells are early cancer progenitor/stem cells that provide a model for patients with poor prognosis. Br J Cancer 98:756–765
61. Prince ME, Sivanandan R, Kaczorowski A et al (2007) Identification of a subpopulation of cells with cancer stem cell properties in head and neck squamous cell carcinoma. Proc Natl Acad Sci USA 104:973–978

62. Ma S, Chan KW, Hu L et al (2007) Identification and characterization of tumorigenic liver cancer stem/progenitor cells. Gastroenterology 132:2542–2556
63. Yang ZF, Ho DW, Ng MN et al (2008) Significance of $CD90^+$ cancer stem cells in human liver cancer. Cancer Cell 13:153–166
64. Ricci-Vitiani L, Lombardi DG, Pilozzi E et al (2007) Identification and expansion of human colon-cancer-initiating cells. Nature 445:111–115
65. O'Brien CA, Pollett A, Gallinger S, Dick JE (2007) A human colon cancer cell capable of initiating tumour growth in immunodeficient mice. Nature 445:106–110
66. Dalerba P, Dylla SJ, Park IK et al (2007) Phenotypic characterization of human colorectal cancer stem cells. Proc Natl Acad Sci USA 104:10158–10163
67. Botchkina IL, Rowehl RA, Rivadeneira DE et al (2009) Phenotypic subpopulations of metastatic colon cancer stem cells: genomic analysis. Cancer Genomics Proteomics 6:19–29
68. Li C, Heidt DG, Dalerba P et al (2007) Identification of pancreatic cancer stem cells. Cancer Res 67:1030–1037
69. Hermann PC, Huber SL, Herrler T et al (2007) Distinct populations of cancer stem cells determine tumor growth and metastatic activity in human pancreatic cancer. Cell Stem Cell 1:313–323
70. Atsumi N, Ishii G, Kojima M, Sanada M, Fujii S, Ochiai A (2008) Podoplanin, a novel marker of tumor-initiating cells in human squamous cell carcinoma A431. Biochem Biophys Res Commun 373:36–41
71. Eramo A, Lotti F, Sette G et al (2008) Identification and expansion of the tumorigenic lung cancer stem cell population. Cell Death Differ 15:504–514
72. Schatton T, Murphy GF, Frank NY et al (2008) Identification of cells initiating human melanomas. Nature 451:345–349
73. Takaishi S, Okumura T, Tu S et al (2009) Identification of gastric cancer stem cells using the cell surface marker CD44. Stem Cells 27:1006–1020
74. Curley MD, Therrien VA, Cummings CL et al (2009) CD133 expression defines a tumor initiating cell population in primary human ovarian cancer. Stem Cells 27:2875–2883
75. Zhang S, Balch C, Chan MW et al (2008) Identification and characterization of ovarian cancer-initiating cells from primary human tumors. Cancer Res 68:4311–4320
76. Alvero AB, Chen R, Fu HH et al (2009) Molecular phenotyping of human ovarian cancer stem cells unravels the mechanisms for repair and chemoresistance. Cell Cycle 8:158–166
77. Weigmann A, Corbeil D, Hellwig A, Huttner WB (1997) Prominin, a novel microvilli-specific polytopic membrane protein of the apical surface of epithelial cells, is targeted to plasmalemmal protrusions of non-epithelial cells. Proc Natl Acad Sci USA 94:12425–12430
78. Yin AH, Miraglia S, Zanjani ED et al (1997) AC133, a novel marker for human hematopoietic stem and progenitor cells. Blood 90:5002–5012
79. Corbeil D, Roper K, Hellwig A et al (2000) The human AC133 hematopoietic stem cell antigen is also expressed in epithelial cells and targeted to plasma membrane protrusions. J Biol Chem 275:5512–5520
80. Maw MA, Corbeil D, Koch J et al (2000) A frameshift mutation in prominin (mouse)-like 1 causes human retinal degeneration. Hum Mol Genet 9:27–34
81. Bourseau-Guilmain E, Griveau A, Benoit JP, Garcion E (2011) The importance of the stem cell marker prominin-1/CD133 in the uptake of transferrin and in iron metabolism in human colon cancer Caco-2 cells. PLoS One 6:e25515
82. Du L, Wang H, He L et al (2008) CD44 is of functional importance for colorectal cancer stem cells. Clin Cancer Res 14:6751–6760
83. Goodison S, Urquidi V, Tarin D (1999) CD44 cell adhesion molecules. Mol Pathol 52:189–196
84. Ahrens T, Assmann V, Fieber C et al (2001) CD44 is the principal mediator of hyaluronic-acid-induced melanoma cell proliferation. J Invest Dermatol 116:93–101
85. Misra S, Toole BP, Ghatak S (2006) Hyaluronan constitutively regulates activation of multiple receptor tyrosine kinases in epithelial and carcinoma cells. J Biol Chem 281:34936–34941
86. Sheridan C, Kishimoto H, Fuchs RK et al (2006) $CD44^+/CD24^-$ breast cancer cells exhibit enhanced invasive properties: an early step necessary for metastasis. Breast Cancer Res 8:R59

87. Omara-Opyene AL, Qiu J, Shah GV, Iczkowski KA (2004) Prostate cancer invasion is influenced more by expression of a CD44 isoform including variant 9 than by Muc18. Lab Invest 84:894–907
88. Zhang T, Huang XH, Dong L et al (2010) PCBP-1 regulates alternative splicing of the CD44 gene and inhibits invasion in human hepatoma cell line HepG2 cells. Mol Cancer 9:72
89. Li Y, Heldin P (2001) Hyaluronan production increases the malignant properties of mesothelioma cells. Br J Cancer 85:600–607
90. Paradis V, Eschwege P, Loric S et al (1998) De novo expression of CD44 in prostate carcinoma is correlated with systemic dissemination of prostate cancer. J Clin Pathol 51:798–802
91. Orian-Rousseau V (2010) CD44, a therapeutic target for metastasising tumours. Eur J Cancer 46:1271–1277
92. Park HY, Lee KJ, Lee SJ, Yoon MY Screening of Peptides Bound to Breast Cancer Stem Cell Specific Surface Marker CD44 by Phage Display. Mol Biotechnol 51:212–220
93. Joshua B, Kaplan MJ, Doweck I et al Frequency of cells expressing CD44, a Head and Neck cancer stem cell marker: correlation with tumor aggressiveness. Head Neck 34:42–49
94. Colnot DR, Roos JC, de Bree R et al (2003) Safety, biodistribution, pharmacokinetics, and immunogenicity of 99mTc-labeled humanized monoclonal antibody BIWA 4 (bivatuzumab) in patients with squamous cell carcinoma of the head and neck. Cancer Immunol Immunother 52:576–582
95. Borjesson PK, Postema EJ, Roos JC et al (2003) Phase I therapy study with (186)Re-labeled humanized monoclonal antibody BIWA 4 (bivatuzumab) in patients with head and neck squamous cell carcinoma. Clin Cancer Res 9(39):61 S–72 S
96. Tijink BM, Buter J, de Bree R et al (2006) A phase I dose escalation study with anti-CD44v6 bivatuzumab mertansine in patients with incurable squamous cell carcinoma of the head and neck or esophagus. Clin Cancer Res 12:6064–6072
97. Jin L, Hope KJ, Zhai Q, Smadja-Joffe F, Dick JE (2006) Targeting of CD44 eradicates human acute myeloid leukemic stem cells. Nat Med 12:1167–1174
98. Rocco A, Liguori E, Pirozzi G et al (2011) CD133 and CD44 cell surface markers do not identify cancer stem cells in primary human gastric tumours. J Cell Physiol 227:2686–2693
99. Eppert K, Takenaka K, Lechman ER et al (2011) Stem cell gene expression programs influence clinical outcome in human leukemia. Nat Med 17:1086–1093
100. Park SY, Lee HE, Li H, Shipitsin M, Gelman R, Polyak K (2010) Heterogeneity for stem cell-related markers according to tumor subtype and histologic stage in breast cancer. Clin Cancer Res 16:876–887
101. Ali HR, Dawson SJ, Blows FM, Provenzano E, Pharoah PD, Caldas C (2011) Cancer stem cell markers in breast cancer: pathological, clinical and prognostic significance. Breast Cancer Res 13:R118
102. Allalunis-Turner MJ, Siemann DW (1986) Recovery of cell subpopulations from human tumour xenografts following dissociation with different enzymes. Br J Cancer 54:615–622
103. Abuzakouk M, Feighery C, O'Farrelly C (1996) Collagenase and Dispase enzymes disrupt lymphocyte surface molecules. J Immunol Methods 194:211–216
104. Reynolds BA, Weiss S (1992) Generation of neurons and astrocytes from isolated cells of the adult mammalian central nervous system. Science 255:1707–1710
105. Yu S, Zhang JZ, Zhao CL, Zhang HY, Xu Q (2004) Isolation and characterization of the $CD133^+$ precursors from the ventricular zone of human fetal brain by magnetic affinity cell sorting. Biotechnol Lett 26:1131–1136
106. Ponti D, Costa A, Zaffaroni N et al (2005) Isolation and in vitro propagation of tumorigenic breast cancer cells with stem/progenitor cell properties. Cancer Res 65:5506–5511
107. Fang D, Nguyen TK, Leishear K et al (2005) A tumorigenic subpopulation with stem cell properties in melanomas. Cancer Res 65:9328–9337
108. Gou S, Liu T, Wang C et al (2007) Establishment of clonal colony-forming assay for propagation of pancreatic cancer cells with stem cell properties. Pancreas 34:429–35

109. Duhagon MA, Hurt EM, Sotelo-Silveira JR, Zhang X, Farrar WL (2010) Genomic profiling of tumor initiating prostatospheres. BMC Genomics 11:324
110. Todaro M, Alea MP, Di Stefano AB et al (2007) Colon cancer stem cells dictate tumor growth and resist cell death by production of interleukin-4. Cell Stem Cell 1:389–402
111. Yuan X, Curtin J, Xiong Y et al (2004) Isolation of cancer stem cells from adult glioblastoma multiforme. Oncogene 23:9392–9400
112. Galli R, Binda E, Orfanelli U et al (2004) Isolation and characterization of tumorigenic, stem-like neural precursors from human glioblastoma. Cancer Res 64:7011–7021
113. Gibbs CP, Kukekov VG, Reith JD et al (2005) Stem-like cells in bone sarcomas: implications for tumorigenesis. Neoplasia 7:967–976
114. Suslov ON, Kukekov VG, Ignatova TN, Steindler DA (2002) Neural stem cell heterogeneity demonstrated by molecular phenotyping of clonal neurospheres. Proc Natl Acad Sci USA 99:14506–14511
115. Patrawala L, Calhoun T, Schneider-Broussard R et al (2006) Highly purified $CD44^+$ prostate cancer cells from xenograft human tumors are enriched in tumorigenic and metastatic progenitor cells. Oncogene 25:1696–1708
116. Jensen JB, Parmar M (2006) Strengths and limitations of the neurosphere culture system. Mol Neurobiol 34:153–161
117. Sladek NE (2003) Human aldehyde dehydrogenases: potential pathological, pharmacological, and toxicological impact. J Biochem Mol Toxicol 17:7–23
118. Moreb JS (2008) Aldehyde dehydrogenase as a marker for stem cells. Curr Stem Cell Res Ther 3:237–246
119. Storms RW, Trujillo AP, Springer JB et al (1999) Isolation of primitive human hematopoietic progenitors on the basis of aldehyde dehydrogenase activity. Proc Natl Acad Sci USA 96:9118–9123
120. Alison MR, Guppy NJ, Lim SM, Nicholson LJ (2010) Finding cancer stem cells: are aldehyde dehydrogenases fit for purpose? J Pathol 222:335–344
121. Pearce DJ, Taussig D, Simpson C et al (2005) Characterization of cells with a high aldehyde dehydrogenase activity from cord blood and acute myeloid leukemia samples. Stem Cells 23:752–760
122. Ginestier C, Hur MH, Charafe-Jauffret E et al (2007) ALDH1 is a marker of normal and malignant human mammary stem cells and a predictor of poor clinical outcome. Cell Stem Cell 1:555–567
123. Boonyaratanakornkit JB, Yue L, Strachan LR et al (2010) Selection of tumorigenic melanoma cells using ALDH. J Invest Dermatol 130:2799–2808
124. van den Hoogen C, van der Horst G, Cheung H et al (2010) High aldehyde dehydrogenase activity identifies tumor-initiating and metastasis-initiating cells in human prostate cancer. Cancer Res 70:5163–5173
125. Silva IA, Bai S, McLean K et al (2011) Aldehyde dehydrogenase in combination with CD133 defines angiogenic ovarian cancer stem cells that portend poor patient survival. Cancer Res 71:3991–4001
126. Liang D, Shi Y (2012) Aldehyde dehydrogenase-1 is a specific marker for stem cells in human lung adenocarcinoma. Med Oncol 29:633–639
127. Wang L, Park P, Zhang H, La Marca F, Lin CY (2011) Prospective identification of tumorigenic osteosarcoma cancer stem cells in OS99-1 cells based on high aldehyde dehydrogenase activity. Int J Cancer 128:294–303
128. Hwang-Verslues WW, Kuo WH, Chang PH et al (2009) Multiple lineages of human breast cancer stem/progenitor cells identified by profiling with stem cell markers. PLoS One 4:e8377
129. Yu C, Yao Z, Dai J et al (2011) ALDH activity indicates increased tumorigenic cells, but not cancer stem cells, in prostate cancer cell lines. In Vivo 25:69–76
130. Zhou J, Wang H, Cannon V, Wolcott KM, Song H, Yates C (2011) Side population rather than CD133(+) cells distinguishes enriched tumorigenicity in hTERT-immortalized primary prostate cancer cells. Mol Cancer 10:112

131. Camargo FD, Chambers SM, Drew E, McNagny KM, Goodell MA (2006) Hematopoietic stem cells do not engraft with absolute efficiencies. Blood 107:501–507
132. Grotenhuis BA, Wijnhoven BP, van Lanschot JJ (2012) Cancer stem cells and their potential implications for the treatment of solid tumors. J Surg Oncol doi:10.1002/jso.23069
133. Jones RJ (2009) Cancer stem cells-clinical relevance. J Mol Med (Berl) 87:1105–1110
134. Zeppernick F, Ahmadi R, Campos B et al (2008) Stem Cell Marker CD133 affects clinical outcome in glioma patients. Clin Cancer Res 14:123–129
135. Liu R, Wang X, Chen GY et al (2007) The prognostic role of a gene signature from tumorigenic breast-cancer cells. N Engl J Med 356:217–226

Chapter 2
DNA Repair Pathways and Mechanisms

Thomas S. Dexheimer

Abstract Our cells are constantly exposed to insults from endogenous and exogenous agents that can introduce damage into our DNA and generate genomic instability. Many of these lesions cause structural damage to DNA and can alter or eliminate fundamental cellular processes, such as DNA replication or transcription. DNA lesions commonly include base and sugar modifications, single- and double-strand breaks, DNA-protein cross-links, and base-free sites. To counteract the harmful effects of DNA damage, cells have developed a specialized DNA repair system, which can be subdivided into several distinct mechanisms based on the type of DNA lesion. These processes include base excision repair, mismatch repair, nucleotide excision repair, and double-strand break repair, which comprise both homologous recombination and non-homologous end-joining. Although a complex set of cellular responses are elicited following DNA damage, this chapter provides an introduction to the specific molecular mechanisms of recognition, removal, and repair of DNA damage.

2.1 Overview

It is estimated that each of the $\sim 10^{13}$ cells within the human body incurs tens of thousands of DNA-damaging events per day [1]. DNA exclusively serves as the repository for the genetic information in each living cell and its integrity and stability are of much greater consequence than other cellular components, such as RNA and proteins. DNA damage can interfere with essential cellular processes, such as transcription or replication, and can compromise the viability of the cell. Specific DNA lesions can also induce mutations that cause cancer or other diseases as well as contribute to the aging process [2]. Thus, cells have evolved a network of DNA repair mechanisms to remove different types of DNA damage. Regardless of the type of lesion and the mechanism required for its repair, cells initiate a highly coordinated cascade of events—collectively known as the DNA damage response (DDR)—that

T. S. Dexheimer (✉)
NIH Chemical Genomics Center, National Center for Advancing Translational Sciences, NIH, 9800 Medical Center Drive, MSC 3370, Bethesda, MD 20892-3370, USA
e-mail: dexheimt@mail.nih.gov

L. A. Mathews et al. (eds.), *DNA Repair of Cancer Stem Cells*,
DOI 10.1007/978-94-007-4590-2_2, © Springer Science+Business Media Dordrecht 2013

senses the DNA damage, signals its presence, and mediates its repair. For example, the DDR may transiently arrest the cell cycle to allow for efficient DNA damage repair prior to replication or mitosis [3, 4] or signal cells to activate apoptosis under circumstances of persistent or irreparable DNA damage [5]. The importance of DDR is underscored by the prevalence of neurological and cancer susceptibility disorders, such as Ataxia-telangiectasia, Fanconi anemia, and Xeroderma pigmentosum, that are caused by DNA repair deficiencies [6]. In this chapter the major types of DNA damage and the respective molecular pathways that function in their repair will be introduced (see Fig. 2.1).

2.2 Types of DNA Damage

As a prelude to the repair of damaged DNA, we must first take into consideration the collection of damage products. DNA, like any other molecule, is subject to chemical reactions. DNA damage may result from either intrinsic or extrinsic agents. In general, the vast majority of DNA modifications are endogenous in origin (for review, see [7]). The simplest form of endogenous DNA damage is spontaneous hydrolysis [8]. The *N*-glycosidic bond between the DNA base and the deoxyribose is particularly prone to acid-catalyzed hydrolysis. Abasic or AP sites (apurinic/apyrimidinic sites), which are the products of hydrolytic nucleobase loss, are estimated to occur at a rate of approximately 10,000 per cell per day [8, 9]. In fact, abasic sites are also created by cellular design during the course of BER (see Sect. 2.3.1). Furthermore, abasic sites are chemically liable and can undergo β-elimination that results in DNA strand scission [10]. Another common reaction involving hydrolysis is the deamination of DNA bases carrying exocyclic amino groups [8, 11]. The most frequent of these lesions is the formation of uracil from cytosine occurring at an estimated 100–500 times per cell per day [12, 13]. Adenine and guanine, may also spontaneously deaminate to form hypoxanthine and xanthine, respectively, although at a much lower rate [14].

DNA is also susceptible to chemical modification by reactive molecules that are created during normal cellular metabolism. Among the most important of these molecules are reactive oxygen species (ROS), which include O_2^-, H_2O_2, and $^\bullet OH$ (for reviews, see [15, 16]). ROS generate over one hundred different oxidative DNA adducts, such as base modification, deoxyribose oxidation, single- or double-strand breakage, and DNA-protein cross-links [17]. Endogenous reactive nitrogen species, primarily nitric oxide ($NO^\bullet$) and its by-products, can also produce similar oxidative adducts [18]. The most extensively studied oxidative DNA lesion is the 8-oxoguanine, which is routinely used as an analytical measure of oxidative DNA damage in biological systems [19]. An additional type of DNA damage related to endogenous reactive molecules is alkylation. The putative candidates of such agents include the endogenous methyl donor, S-adenosylmethionine (SAM), nitrosated amines, and methyl radicals generated by lipid peroxidation [7, 20]. The primary sites of alkylation are the O^- and N-atoms of nucleobases.

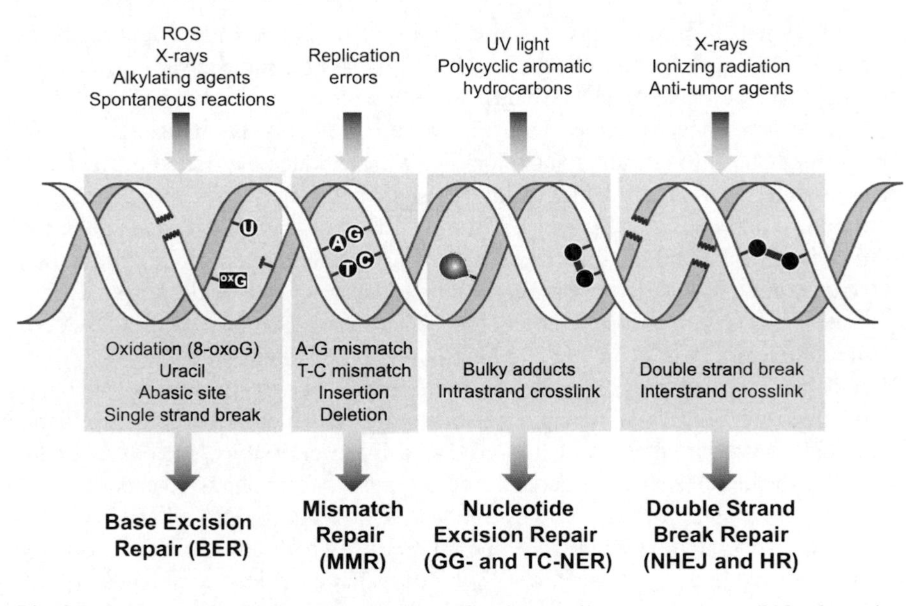

Fig. 2.1 DNA damage and repair mechanisms. The diagram illustrates common DNA damaging agents, examples of DNA lesions caused by these agents, and the relevant DNA repair mechanism responsible for their removal. (Figure adapted from [83])

Endogenous genomic damage can also arise due to unavoidable errors resulting from physiological DNA processing reactions. For example, DNA mismatches as well as insertions and deletions are occasionally introduced (at a rate of 10^{-4} to 10^{-6}) as a result of misincorporation of bases by replicative DNA polymerases [21]. At the same time, erroneous incorporation of chemically altered nucleotide precursors, such as 8-oxo-dGTP and dUTP [22], also represents a significant source of replication-related DNA damage. In addition, abortive topoisomerase activity yields an irregular type of lesion wherein DNA strand breaks feature covalent linkage of the enzyme to the DNA termini [23, 24]. Likewise, the DNA repair processes themselves may also be error prone and introduce supplemental DNA damage [25].

Besides the numerous endogenous sources of DNA damage, cellular DNA is also under constant attack from exogenous or environmental DNA-damaging agents. These include physical stresses, such as ultraviolet light (UV) from the sun, which primarily causes two types of DNA lesions, namely cyclobutane pyrimidine dimers and 6–4 pyrimidone photoproducts, both of which consist of an atypical covalent bond between adjacent pyrimidine bases [26]. Another external, physical source of DNA damage is ionizing radiation, which can originate from both natural (e.g., cosmic and gamma radiation) and artificial sources (e.g., medical treatments, such as X-rays and radiotherapy). Ionizing radiation induces a variety of DNA lesions, the most harmful of these being double-strand breaks. DNA can also incur damage indirectly from ionizing radiation through the production of ROS [27].

In addition to the physical insults, the cell must also contend with several chemical sources of DNA damage (for reviews, see [28, 29]). For example, a variety of chemical agents (i.e., clinical drugs) have been developed over the years to target DNA as a means to treat cancer or other diseases. These include alkylating agents, such as methyl methanesulfonate and temozolomide, which induce alkylation of the DNA bases as well as bifunctional alkylating agents, such as nitrogen mustards, platinum compounds, and the natural product mitomycin C, that cause DNA damage in the form of intrastrand and interstrand cross-links [30]. Chemotherapeutic drugs, such as topoisomerse I or II inhibitors (e.g., camptothecin or etoposide, respectively), generate single-strand or double-strand breaks by trapping topoisomerase–DNA covalent complexes, respectively [31]. Other well-studied environmentally occurring DNA-damaging chemicals include N-nitrosoamines, heterocyclic amines, and polycyclic aromatic hydrocarbons (e.g., benzo[*a*]pyrene), which are commonly found in the diet, with the latter also being produced in air emissions, such as cigarette smoke and vehicle exhaust. In general, these types of compounds covalently bond to various sites on the DNA bases to form the so-called bulky DNA adducts. Similar adducts are generated between DNA and aflatoxins, which are naturally occurring toxins produced by fungi in the genus *Aspergillus* that grow in several types of food crops [32].

2.3 DNA Repair Mechanisms

To compensate for the many types of DNA damage that occur, cells have developed multiple repair mechanisms wherein each corrects a different subset of lesions. At a minimum, most would agree that mammalian cells utilize five major DNA repair mechanisms: base excision repair (BER), mismatch repair (MMR), nucleotide excision repair (NER), and double-strand break repair, which includes both homologous recombination (HR) and non-homologous end joining (NHEJ) (for comprehensive review, see [33]). For reference, Table 2.1 outlines specific genes that are associated with each DNA repair mechanism.

2.3.1 Base Excision Repair (BER)

BER, as the name implies, is the predominant mechanism responsible for the repair of damaged DNA bases that, in contrast to NER (see Sect. 2.3.3), do not significantly distort the overall structure of the DNA helix (for detailed review of BER, see [34]). BER is described as a highly coordinated pathway of consecutive enzymatic reactions. However, several distinct BER sub-pathways occur, which are contingent on the type of damage encountered at the onset as well as throughout the BER process. BER is typically initiated by the series of lesion-specific DNA glycosylases that remove the damaged base by cleaving the *N*-glycosidic bond linking the base to its corresponding deoxyribose, leading to the production of an AP or abasic

Table 2.1 Essential genes of the five major DNA repair mechanisms

Base excision repair (BER)	DNA glycosylase, APE1, XRCC1, PNKP, Tdp1, APTX, DNA polymerase β, FEN1, DNA polymerase δ or ε, PCNA-RFC, PARP
Mismatch repair (MMR)	MutSα (MSH2-MSH6), MutSβ (MSH2-MSH3), MutLα (MLH1-PMS2), MutLβ (MLH1-PMS2), MutLγ (MLH1-MLH3), Exo1, PCNA-RFC
Nucleotide excision repair (NER)	XPC-Rad23B-CEN2, UV-DDB (DDB1-XPE), CSA, CSB, TFIIH, XPB, XPD, XPA, RPA, XPG, ERCC1-XPF, DNA polymerase δ or ε
Homologous recombination (HR)	Mre11-Rad50-Nbs1, CtIP, RPA, Rad51, Rad52, BRCA1, BRCA2, Exo1, BLM-TopIIIα, GEN1-Yen1, Slx1-Slx4, Mus81/Eme1
Non-homologous end-joining (NHEJ)	Ku70-Ku80, DNA-PKc, XRCC4-DNA ligase IV, XLF

site. At least twelve DNA glycosylases have been identified to date, each acting upon a single or small number of partially overlapping base lesions [35]. Despite their structural diversity, all DNA glycosylases utilize a base-flipping mechanism in which the target base is 'flipped' to an extra helical position for excision from DNA [36]. The resultant AP site is both an intermediate product of BER and a highly prevalent DNA lesion produced by spontaneous base loss. In either case, AP sites are generally repaired by apurinc/apyrmidinic endonuclease 1 (APE1), the second enzyme in the canonical BER pathway. APE1 hydrolyzes the phosphodiester backbone immediately 5′ to the AP site, creating a single-strand break flanked by 3′-OH and 5′-deoxyribose phosphate (5′-dRP) termini [37]. Alternatively, some DNA glycosylases have an associated AP lyase activity and are also capable of cleaving AP sites via a β-elimination reaction to produce 3′-phospho-α, β-unsaturated aldehyde and 5′-phosphate at the margins of the break. A subset of these bifunctional enzymes, such as the oxidized base-specific DNA glycosylase/lyases NEIL1 and NEIL2, catalyze successive β- and δ-elimination converting the 3′-phospho-α, β-unsaturated aldehyde to a 3′-phosphate.

Regardless of mechanism, incision of the phosphodiester bond results in a BER intermediate strand break harboring 3′- and 5′-blocking lesions. To allow completion of the repair process, these blocked termini must be restored to conventional 3′-OH and 5′-phosphate ends, which are essential for DNA polymerase and subsequent DNA ligase reactions. Different DNA end-processing enzymes carry out the removal of these abnormal ends depending on whether cleavage occurred 3′ or 5′ to the AP site. APE1, for example, in addition to its major AP endonuclease activity also has intrinsic 3′-phosphodiesterase activity, permitting restoration of 3′-OH from 3′-phospho-α, β-unsaturated aldehyde. The 3′-phophate product that is generated by specific bifunctional DNA glycosylases is converted to a 3′-OH by the 3′-phosphatase activity of PNKP (polynucleotide kinase 3′-phosphatase). Conversely, removal of the 5′-dRP occurs following template-guided gap filling by DNA polymerase β via its associated dRP lyase activity.

Besides the scheduled DNA single-strand breaks that arise as BER intermediates, numerous involuntary DNA single-strand breaks can also occur both through direct and indirect mechanisms (see Sect. 2.2). Such single-strand interruptions are processed and repaired by many of the same enzymes that are responsible for the later stages of BER. The termini of most, if not all, single-strand breaks contain 3′- and/or 5′-blocking lesions. For example, the most common blocking lesions at ROS-induced DNA strand breaks are 3′-phosphoglycolate and 3′-phosphoglycolaldehyde, which are generally processed by the 3′-phosphodesterase activity of APE1, or 3′ phosphate, which is removed by PNKP. Tyrosyl-DNA phosphodiesterase 1 (Tdp1) is an end-processing enzyme that repairs several 3′-blocking termini including 3′-phosphoglycolate; however, its preferred substrate is the 3′-phosphotyrosyl bond, which stems from dead-end topoisomerase I reactions [38]. Likewise, aprataxin (APTX) is another specific end-processing enzyme, which specifically repairs abortive 5′-adenylate intermediates of DNA ligase activity [39]. Thus, DNA end-processing is perhaps the most diverse, yet often redundant, enzymatic step of BER, largely due to the broad range of termini that can be generated [40].

The next steps in the BER process involve repair of the DNA strand break through DNA synthesis and ligation. The synthesis/ligation step is divided into two sub-pathways, short-patch and long patch BER, based on whether a single or several nucleotides are incorporated at the DNA strand break site, respectively [41]. The paradigm for short-patch BER encompasses single nucleotide gap filling and removal of the 5′-dRP by DNA polymerase β and successive ligation of the DNA ends by either DNA ligase I or the complex of DNA ligase III and XRCC1. Short-patch BER represents approximately 80–90 % of all BER. Long-patch BER is normally only initiated as a result of 5′-blocking lesions that are refractory to DNA polymerase β lyase activity. Long-patch BER demands several proteins associated with DNA replication, including DNA polymerase δ or ε, PCNA (proliferating cell nuclear antigen), RFC (replication factor-C), FEN1 (flap endonuclease-1), and DNA ligase I. Specifically, DNA polymerase β, δ, or ε accompanied by PCNA elongate the 3′-OH into the repair gap and displace the 5′-lesion as part of a DNA fragment or 'flap' oligonucleotide. The flap structure is then removed by FEN1 and DNA ligase I sequentially seals the nick that has been relocated downstream of the original nucleotide damage site.

In addition to the factors mentioned above, there are secondary proteins that are known to play a facilitative role in BER. Most notably among these are X-ray repair cross-complementing protein 1 (XRCC1) and poly (ADP-ribose) polymerase 1 (PARP1). XRCC1 has no known enzymatic activity, but rather functions as a molecular scaffold that orchestrates the assembly of several enzymatic components involved in the BER process. For instance, XRCC1 has been shown to interact with several BER proteins, including multiple DNA glycosylases, DNA polymerase β, APE1, ligase III, PNKP, Tdp1, and APTX [42]. Although no catalytic function has been ascribed to XRCC1, direct binding to nicked and gapped DNA has been demonstrated via its N-terminal domain [43]. Additionally, PARP-1 also physically interacts with XRCC1. PARP1 is an abundant nuclear protein that acts as a molecular sensor of DNA strand breaks. Upon binding to its DNA target, PARP-1 catalyzes the poly (ADP-ribosyl)ation (PAR) of itself, in addition to several other protein substrates. Once formed, this PAR modification allows for recruitment of repair proteins, such as XRCC1. At the same time, the dense negative charge of PAR results in the release of PARP-1 from DNA, which permits access of repair proteins to the DNA damage site [44]. Overall, BER is a multistep process that requires the sequential activity of several proteins and consists of numerous entry points based on the type of damage encountered.

2.3.2 Mismatch Repair (MMR)

The MMR system plays an essential role in post-replication repair of misincorporated bases that have escaped the proofreading activity of replication polymerases. In addition to mismatched bases, MMR proteins also correct insertion/deletion loops (IDLs) that result from polymerase slippage during replication of repetitive DNA sequences. The significance of this pathway is corroborated by the fact that MMR deficient cells are said to display a mutator phenotype, which is characterized by invariably microsatellite instability and an elevated mutation frequency. More importantly, germline mutations in MMR genes are predisposed to a variety of cancers, including hereditary non-polyposis colon cancer, also known as Lynch syndrome [45]. The MMR pathway can be divided into three principle steps: a recognition step where mispaired bases are recognized, an excision step where the error-containing strand is degraded resulting in a gap, and a repair synthesis step, where the gap is filled by the DNA resynthesis (for detailed reviews of MMR, see [46–48]).

The MMR process is highly conserved from *E.coli* to humans. The canonical human MMR pathway is carried out by two major protein complexes, which are so-called MutS and MutL, based on their homology to the *E.coli* MMR proteins [49]. While MutS is responsible for mismatch recognition, MutL couples the recognition of the mispaired bases by the MutS complexes to downstream MMR events, which lead to the removal of the strand containing the error. In mammalians, the initial mismatch recognition step is fulfilled by two MutS activities that function as heterodimers. The MSH2-MSH6 heterodimer, also known as MutSα, preferentially recognizes base-base mismatches and small IDLs of one or two nucleotides, while MutSβ, the heterodimer of MSH2 and MSH3 recognizes larger IDLs. Formation of the MutS-DNA complex is followed by ATP-dependent recruitment of MutL homolog (MLH) complexes. Three MutL activities have been identified and, like MutS, also function as heterodimeric complexes. MutLα, a heterodimer of MLH1 and PMS2, which contains the primary MutL activity ($\sim$90 %) in humans and supports the repair initiated by both MutSα and MutSβ. The two additional MutL heterodimers consist of MLH1/PMS2 (MutLβ) and MLH1/MLH3 (MutLγ), which may play minor roles in MMR.

Assembly of the ATP-dependent MutS-MutL-DNA heteroduplex ternary complex is necessary to activate exonuclease mediated degradation of the error-containing strand [50]. In humans, this degradation is performed by exonuclease 1 (Exo1) through its $5'$ to $3'$ exonucleolytic activity [51]. The entry point for Exo1, which may be thousands of nucleotides from the mismatch, is generated via single-strand scission by the PCNA/replication factor C (RFC)-dependent endonuclease activity of MutLα [52]. The extensive gap left by Exo1 is then resynthesized by DNA polymerase δ, which is accompanied by at least two other proteins, PCNA and replication protein A (RPA). Lastly, MMR is completed by DNA ligase I sealing of the remaining nick.

2.3.3 Nucleotide Excision Repair (NER)

NER is a highly versatile repair pathway that can recognize and remove a wide variety of bulky, helix-distorting lesions from DNA. The most significant of these lesions are pyrimidine dimers, such as cyclobutane pyrimidine dimers (CPD) and 6–4 photoproducts, which are produced by the UV component of sunlight. Another noteworthy substrate of NER is cisplatin-DNA intrastrand crosslinks. NER is mediated by the sequential assembly of repair proteins at the site of the DNA lesion. While mechanistically similar to BER, the NER pathway is more complex, requiring some thirty different proteins to carry out a multi-step 'cut-and-patch'-like mechanism. These steps involve DNA damage recognition, local opening of the DNA helix around the lesion, excision of a short single-strand segment of DNA spanning the lesion, and sequential repair synthesis and strand ligation (for detailed reviews of NER, see [53–55]). The biological importance of NER is supported by the fact that defects in NER cause several human genetic disorders, including xeroderma pigmentosum, Cockayne syndrome, and trichothiodystrophy, which are all characterized by extreme sun sensitivity. In addition, these diseases demonstrate overlapping symptoms associated with cancer, developmental delay, immunological defects, neurodegeneration, and premature aging [56, 57].

The NER system consists of two related subpathways, termed global genome NER (GG-NER) and transcription-coupled NER (TC-NER). As the names imply, GG-NER eliminates DNA lesions throughout the genome, while TC-NER is preferentially responsible for repairing lesions located on the coding strand of actively transcribed genes. Both pathways are mechanistically the same, apart from the initial damage recognition step. In GG-NER, the principle damage recognition factor is the XPC/HR23B/CEN2 (XP complementation group C/Rad23 homolog B/Centrin-2) protein complex [58]. HR23B and CEN2 are accessory proteins that increase both the affinity and specificity of XPC binding to helix-distorting DNA damage. In addition, the DNA binding affinity of XPC generally correlates with the degree of helical distortion [59]. For example, XPC has low affinity to lesions that are caused by only minor distortions, such as UV-induced CPDs. Thus, an auxiliary damage-recognizing complex called the UV-damaged DNA binding complex (UV-DDB), which consists of two subunits, DDB1 and XPE (DDB2), initially detects these types of lesions. The binding of UV-DDB to damaged DNA induces an increase in helix distortion (i.e., DNA bending), which subsequently facilitates the recruitment of the XPC complex to the damage site [60]. In contrast, damage recognition in TC-NER is initiated when an elongating RNA polymerase II (RNAPII) is arrested upon encountering a site of DNA damage [61]. Subsequently, two TC-NER-specific proteins, Cockayne syndrome A (CSA) and B (CSB), are thought to displace the stalled RNAPII to allow NER proteins access to the lesion [62].

Following damage recognition, both GG-NER and TC-NER proceed through the common 'core' NER reactions. Initially, either the XPC complex in GG-NER or, presumably, CSB and CSA in TC-NER recruit the multi-subunit (ten protein complex) and the multi-functional transcription factor TFIIH to the site of damage. Next, two TFIIH-associated, ATP-dependent helicases XPB and XPD orchestrate the asymmetric unwinding of the DNA helix to form a ~30 nucleotide bubble flanking the lesion. Initial unwinding permits access of XPA to the damaged region, which provides a second level of damage recognition in addition to ensuring that undamaged DNA is not subjected to excision repair. The binding of XPA is accompanied by the heterotrimeric, single stranded DNA binding protein RPA (replication protein A), which allows for complete extension and subsequent stabilization of the so-called pre-incision complex. In the subsequent step, two structure-specific endonucleases XPG and XPF/ERCC1 cleave the DNA at positions 3′ and 5′ to the damage, respectively, leading to excision of the lesion-containing oligonucleotide of about 30 nucleotides. Lastly, DNA polymerase δ or ε uses the undamaged strand as a template to resynthesize the resulting gap. The nick of the repaired strand is then sealed by DNA ligase, thus completing the NER process.

2.3.4 Double-Strand Break Repair

Double-strand breaks (DSBs) are amongst the most biologically hazardous types of DNA damage. For instance, a single unrepaired DSB is often sufficient to cause cell death. In addition, inaccurate repair can lead to deletions or chromosomal aberrations, events that associated with the development of cancer or other genomic instability syndromes. Thus, the repair of DSBs is both critical for cell survival and maintenance of genome integrity [63, 64]. The two main mechanisms by which mammalian cells repair DSBs are homologous recombination (HR) and non-homologous end-joining (NHEJ). These two repair systems differ in their requirement for a homologous template DNA and in the fidelity of DSB repair. HR-directed repair is largely an error-free mechanism as it utilizes the genetic information contained in the undamaged sister chromatid as a template (for review, see [65]). In contrast, NHEJ is normally error-prone and involves elimination of DSBs by direct ligation of the broken ends (for review, see [66]). NHEJ is reasoned to be the predominant pathway in mammalian cells operating in all phases of the cell cycle, while HR is restricted to the late-S and G2 phases. The basic mechanisms of these pathways and the factors involved are briefly outlined below.

2.3.4.1 Homologous Recombination (HR)

Much of our current knowledge concerning the mechanism of eukaryotic homology-directed repair is contributed to studies in bacteria and yeast, where HR is most efficient. HR can be conceptually divided into three phases: presynapsis, synapsis, and postsynapsis. During presynapsis, the DNA ends surrounding the DSB are processed through 5′ to 3′ end resection to generate molecules with 3′-single-stranded tails. The heterotrimeric MRN complex (Mre11-Rad50-Nbs1) together with CtIP (RBBP8) are responsible for the initiation of resection in which the 5′-ends on either side of the DSB are trimmed back to create short 3′-overhangs of single-strand DNA [67]. The second step in the 5′ to 3′ resection is presumably continued by the combined action of BLM helicase (Bloom syndrome, RecQ helicase-like) and Exo1 exonuclease [68]. Following end resection, single-stranded DNA tails are bound by RPA to remove disruptive secondary structures that would otherwise obstruct binding of Rad51 recombinase. RPA is subsequently replaced by Rad51 in conjunction with several mediator proteins, such as Rad52, BRCA2, and a group of proteins known as Rad51 paralogs (RAD51B, RAD51C, RAD51D, XRCC2, and XRCC3) [69]. The Rad51-coated single-stranded DNA tail, also referred to as the Rad51 nucleoprotein filament, then executes the DNA sequence homology search, which is the central reaction of HR. Once the homologous DNA has been identified, Rad51 mediates DNA strand invasion reaction, wherein the damaged DNA strand invades the template DNA duplex (i.e., sister chromatid). Next, DNA synthesis from the 3′-end of the invading strand is carried out by DNA polymerase η followed by successive ligation by DNA ligase I to yield a four-way junction intermediate structure known as a Holliday junction [70]. This recombination intermediate is resolved in one of three ways, by 'dissolution' mediated by the BLM-TopIIIα complex, by symmetrical cleavage by GEN1/Yen1 or Slx1/Slx4, or by asymmetric cleavage by the structure-specific endonuclease Mus81/Eme1 [71–73], resulting in the error-free correction of the DSB.

2.3.4.2 Non-Homologous End-Joining (NHEJ)

The molecular mechanism of NHEJ is mediated by a relatively small number of essential factors that are sequentially recruited to DSB sites. The initial step in the NHEJ process entails recognition and binding of the Ku70/Ku80 heterodimer (Ku) to the exposed DNA termini of the DSB. Structurally, Ku adopts a preformed ring-shaped structure that completely encircles the DNA duplex [74]. Upon binding to DNA, the Ku-DNA complex recruits the catalytic subunit of DNA-dependent protein kinase (DNA-PKcs) to generate the so-called DNA-PK holoenzyme, which exhibits protein kinase activity. The recruitment of DNA-PKcs induces an inward translocation of Ku along the DNA, allowing DNA-PKcs to contact DNA termini [75]. More importantly, the binding of the DNA-PKcs molecules on opposing DSB ends promotes synapsis or tethering of the two DNA molecules. Synapsis of DNA-PKcs also results in autophosphorylation of DNA-PKcs, which allows the DNA termini to become accessible [76]. Like most DNA repair processes, depending on the type and complexity of the DSB break, DNA ends may require modification prior to ligation. For example, DNA termini containing single-stranded overhangs can be made ligatable through either DNA polymerase-mediated fill-in or nucleolytic resection.

The resynthesis of missing nucleotides during NHEJ has been associated with two members of the X family DNA polymerases, Pol μ and Pol λ [77]. Alternatively, the NHEJ-specific nuclease Artemis, whose activities include a DNA-PK independent 5′ to 3′ exonuclease activity as well as a DNA-PK dependent endonuclease activity, which is acquired through phosphorylation by DNA-PK, can excise single-stranded overhangs [78]. Other candidates that may also participate in DNA end 'cleaning' process include several of the lesion-specific BER enzymes, such as APE1, Tdp1, and PNKP [79] (see above), as well as the two functional exonucleases Exo1 and WRN, which is mutated in Werner syndrome patients [80, 81]. Consequently, the same enzymes that participate in the end-processing step of NHEJ are considered to be responsible for the overhang mispairing and the gain or loss of nucleotides associated with NHEJ-mediated repair. After appropriate (or sometimes inappropriate) processing of the DNA termini, ligation of the DNA ends is carried out by DNA ligase IV in conjunction with its binding partner XRCC4. An additional factor, XLF (XRCC4-like factor), interacts with the XRCC4-DNA ligase IV complex to promote DNA ligation [82].

2.4 Conclusion

The biological significance of DNA repair mechanisms is underscored by the fact that their deregulation can contribute to the initiation and progression of cancer. On the other hand, DNA repair can confer resistance to front line cancer treatments (i.e. chemotherapy and radiation), which rely on the generation of DNA damage to kill cancer cells. Thus, the sensitivity of cancer cells to DNA damaging agents is most likely related to intrinsic deficiencies in DNA repair mechanisms. The capacity of cancer cells (or cancer stem cells) to recognize DNA damage and initiate DNA repair is a key mechanism for therapeutic resistance or recurrence. The following chapters will discuss the DNA repair mechanisms that ensure protection of cancer stem cells.

References

1. Lindahl T, Barnes DE (2000) Repair of endogenous DNA damage. Cold Spring Harb Symp Quant Biol 65:127–133
2. Hoeijmakers JH (2009) DNA damage, aging, and cancer. N Engl J Med 361(15):1475–1485
3. Stracker TH, Usui T, Petrini JH (2009) Taking the time to make important decisions: the checkpoint effector kinases Chk1 and Chk2 and the DNA damage response. DNA Repair (Amst) 8(9):1047–1054
4. Zhou BB, Elledge SJ (2000) The DNA damage response: putting checkpoints in perspective. Nature 408(6811):433–439
5. Rich T, Allen RL, Wyllie AH (2000) Defying death after DNA damage. Nature 407(6805):777–783
6. McKinnon PJ (2009) DNA repair deficiency and neurological disease. Nat Rev Neurosci 10(2):100–112
7. De Bont R, van Larebeke N (2004) Endogenous DNA damage in humans: a review of quantitative data. Mutagenesis 19(3):169–185

8. Lindahl T (1993) Instability and decay of the primary structure of DNA. Nature 362(6422):709–715
9. Lindahl T, Nyberg B (1972) Rate of depurination of native deoxyribonucleic acid. Biochemistry 11(19):3610–3618
10. Sugiyama H, Fujiwara T, Ura A et al (1994) Chemistry of thermal degradation of abasic sites in DNA. Mechanistic investigation on thermal DNA strand cleavage of alkylated DNA. Chem Res Toxicol 7(5):673–683
11. Yonekura S, Nakamura N, Yonei S, Zhang-Akiyama QM (2009) Generation, biological consequences and repair mechanisms of cytosine deamination in DNA. J Radiat Res (Tokyo) 50(1):19–26
12. Frederico LA, Kunkel TA, Shaw BR (1990) A sensitive genetic assay for the detection of cytosine deamination: determination of rate constants and the activation energy. Biochemistry 29(10):2532–2537
13. Krokan HE, Drablos F, Slupphaug G (2002) Uracil in DNA–occurrence, consequences and repair. Oncogene 21(58):8935–8948
14. Kow YW (2002) Repair of deaminated bases in DNA. Free Radic Biol Med 33(7):886–893
15. Apel K, Hirt H (2004) Reactive oxygen species: metabolism, oxidative stress, and signal transduction. Annu Rev Plant Biol 55:373–399
16. Marnett LJ (2000) Oxyradicals and DNA damage. Carcinogenesis 21(3):361–370
17. Cadet J, Berger M, Douki T, Ravanat JL (1997) Oxidative damage to DNA: formation, measurement, and biological significance. Rev Physiol Biochem Pharmacol 131:1–87
18. Burney S, Caulfield JL, Niles JC, Wishnok JS, Tannenbaum SR (1999) The chemistry of DNA damage from nitric oxide and peroxynitrite. Mutat Res 424(1–2):37–49
19. Ravanat J-L (2005) Measuring oxidized DNA lesions as biomarkers of oxidative stress: an analytical challenge FABAD. J Pharm Sci 30(2):100–113
20. Major GN, Collier JD (1998) Repair of DNA lesion O6-methylguanine in hepatocellular carcinogenesis. J Hepatobiliary Pancreat Surg 5(4):355–366
21. McCulloch SD, Kunkel TA (2008) The fidelity of DNA synthesis by eukaryotic replicative and translesion synthesis polymerases. Cell Res 18(1):148–161
22. Shimizu M, Gruz P, Kamiya H et al (2003) Erroneous incorporation of oxidized DNA precursors by Y-family DNA polymerases. EMBO Rep 4(3):269–273
23. McClendon AK, Osheroff N (2007) DNA topoisomerase II, genotoxicity, and cancer. Mutat Res 623(1–2):83–97
24. Pourquier P, Pommier Y (2001) Topoisomerase I-mediated DNA damage. Adv Cancer Res 80:189–216
25. Bridges BA (2005) Error-prone DNA repair and translesion synthesis: focus on the replication fork. DNA Repair (Amst) 4(5):618–619, 634
26. Ravanat JL, Douki T, Cadet J (2001) Direct and indirect effects of UV radiation on DNA and its components. J Photochem Photobiol B 63(1–3):88–102
27. Ward JF (1988) DNA damage produced by ionizing radiation in mammalian cells: identities, mechanisms of formation, and reparability. Prog Nucleic Acid Res Mol Biol 35:95–125
28. Wogan GN, Hecht SS, Felton JS, Conney AH, Loeb LA (2004) Environmental and chemical carcinogenesis. Semin Cancer Biol 14(6):473–486
29. Irigaray P, Belpomme D (2010) Basic properties and molecular mechanisms of exogenous chemical carcinogens. Carcinogenesis 31(2):135–148
30. Noll DM, Mason TM, Miller PS (2006) Formation and repair of interstrand cross-links in DNA. Chem Rev 106(2):277–301
31. Sinha BK (1995) Topoisomerase inhibitors. A review of their therapeutic potential in cancer. Drugs 49(1):11–19
32. Bedard LL, Massey TE (2006) Aflatoxin B1-induced DNA damage and its repair. Cancer Lett 241(2):174–183
33. Altieri F, Grillo C, Maceroni M, Chichiarelli S (2008) DNA damage and repair: from molecular mechanisms to health implications. Antioxid Redox Signal 10(5):891–937
34. Zharkov DO (2008) Base excision DNA repair. Cell Mol Life Sci 65(10):1544–1565

35. Jacobs AL, Schar P (2012) DNA glycosylases: in DNA repair and beyond. Chromosoma 121(1):1–20
36. Hitomi K, Iwai S, Tainer JA (2007) The intricate structural chemistry of base excision repair machinery: implications for DNA damage recognition, removal, and repair. DNA Repair (Amst) 6(4):410–428
37. Abbotts R, Madhusudan S (2010) Human AP endonuclease 1 (APE1): from mechanistic insights to druggable target in cancer. Cancer Treat Rev 36(5):425–435
38. Interthal H, Chen HJ, Champoux JJ (2005) Human Tdp1 cleaves a broad spectrum of substrates, including phosphoamide linkages. J Biol Chem 280(43):36518–36528
39. Ahel I, Rass U, El-Khamisy SF et al (2006) The neurodegenerative disease protein aprataxin resolves abortive DNA ligation intermediates. Nature 443(7112):713–716
40. Caldecott KW (2008) Single-strand break repair and genetic disease. Nat Rev Genet 9(8):619–631
41. Fortini P, Dogliotti E (2007) Base damage and single-strand break repair: mechanisms and functional significance of short- and long-patch repair subpathways. DNA Repair (Amst) 6(4):398–409
42. Caldecott KW (2003) XRCC1 and DNA strand break repair. DNA Repair (Amst) 2(9):955–969
43. Marintchev A, Mullen MA, Maciejewski MW, Pan B, Gryk MR, Mullen GP (1999) Solution structure of the single-strand break repair protein XRCC1 N-terminal domain. Nat Struct Biol 6(9):884–893
44. Malanga M, Althaus FR (2005) The role of poly(ADP-ribose) in the DNA damage signaling network. Biochem Cell Biol 83(3):354–364
45. Peltomaki P (2001) Deficient DNA mismatch repair: a common etiologic factor for colon cancer. Hum Mol Genet 10(7):735–740
46. Li GM (2008) Mechanisms and functions of DNA mismatch repair. Cell Res 18(1):85–98
47. Fukui K (2010) DNA mismatch repair in eukaryotes and bacteria. J Nucleic Acids 2010:1–6
48. Larrea AA, Lujan SA, Kunkel TA (2010) SnapShot: DNA mismatch repair. Cell 141(4):730 e1
49. Modrich P (2006) Mechanisms in eukaryotic mismatch repair. J Biol Chem 281(41):30305–30309
50. Galio L, Bouquet C, Brooks P (1999) ATP hydrolysis-dependent formation of a dynamic ternary nucleoprotein complex with MutS and MutL. Nucleic Acids Res 27(11):2325–2331
51. Tran PT, Erdeniz N, Symington LS, Liskay RM (2004) EXO1-A multi-tasking eukaryotic nuclease. DNA Repair (Amst) 3(12):1549–1559
52. Kadyrov FA, Holmes SF, Arana ME et al (2007) Saccharomyces cerevisiae MutLalpha is a mismatch repair endonuclease. J Biol Chem 282(51):37181–37190
53. Shuck SC, Short EA, Turchi JJ (2008) Eukaryotic nucleotide excision repair: from understanding mechanisms to influencing biology. Cell Res 18(1):64–72
54. Costa RM, Chigancas V, Galhardo Rda S, Carvalho H, Menck CF (2003) The eukaryotic nucleotide excision repair pathway. Biochimie 85(11):1083–1099
55. Nouspikel T (2008) Nucleotide excision repair and neurological diseases. DNA Repair (Amst) 7(7):1155–1167
56. Cleaver JE, Lam ET, Revet I (2009) Disorders of nucleotide excision repair: the genetic and molecular basis of heterogeneity. Nat Rev Genet 10(11):756–768
57. Vermeulen W, de Boer J, Citterio E et al (1997) Mammalian nucleotide excision repair and syndromes. Biochem Soc Trans 25(1):309–315
58. Sugasawa K, Ng JM, Masutani C et al (1998) Xeroderma pigmentosum group C protein complex is the initiator of global genome nucleotide excision repair. Mol Cell 2(2):223–232
59. Sugasawa K (2008) XPC: its product and biological roles. Adv Exp Med Biol 637:47–56
60. Sugasawa K (2010) Regulation of damage recognition in mammalian global genomic nucleotide excision repair. Mutat Res 685(1–2):29–37
61. Fousteri M, Mullenders LH (2008) Transcription-coupled nucleotide excision repair in mammalian cells: molecular mechanisms and biological effects. Cell Res 18(1):73–84
62. Hanawalt PC, Spivak G (2008) Transcription-coupled DNA repair: two decades of progress and surprises. Nat Rev Mol Cell Biol 9(12):958–970

63. van Gent DC, Hoeijmakers JH, Kanaar R (2001) Chromosomal stability and the DNA double-stranded break connection. Nat Rev Genet 2(3):196–206
64. Khanna KK, Jackson SP (2001) DNA double-strand breaks: signaling, repair and the cancer connection. Nat Genet 27(3):247–254
65. Li X, Heyer WD (2008) Homologous recombination in DNA repair and DNA damage tolerance. Cell Res 18(1):99–113
66. Lieber MR (2010) The mechanism of double-strand DNA break repair by the nonhomologous DNA end-joining pathway. Annu Rev Biochem 79:181–211
67. Sartori AA, Lukas C, Coates J et al (2007) Human CtIP promotes DNA end resection. Nature 450(7169):509–514
68. Nimonkar AV, Ozsoy AZ, Genschel J, Modrich P, Kowalczykowski SC (2008) Human exonuclease 1 and BLM helicase interact to resect DNA and initiate DNA repair. Proc Natl Acad Sci USA 105(44):16906–16911
69. Forget AL, Kowalczykowski SC (2010) Single-molecule imaging brings Rad51 nucleoprotein filaments into focus. Trends Cell Biol 20(5):269–276
70. McIlwraith MJ, Vaisman A, Liu Y, Fanning E, Woodgate R, West SC (2005) Human DNA polymerase eta promotes DNA synthesis from strand invasion intermediates of homologous recombination. Mol Cell 20(5):783–792
71. Ip SC, Rass U, Blanco MG, Flynn HR, Skehel JM, West SC (2008) Identification of Holliday junction resolvases from humans and yeast. Nature 456(7220):357–361
72. Mimitou EP, Symington LS (2009) Nucleases and helicases take center stage in homologous recombination. Trends Biochem Sci 34(5):264–272
73. Seki M, Nakagawa T, Seki T et al (2006) Bloom helicase and DNA topoisomerase III alpha are involved in the dissolution of sister chromatids. Mol Cell Biol 26(16):6299–6307
74. Walker JR, Corpina RA, Goldberg J (2001) Structure of the Ku heterodimer bound to DNA and its implications for double-strand break repair. Nature 412(6847):607–614
75. Yoo S, Dynan WS (1999) Geometry of a complex formed by double strand break repair proteins at a single DNA end: recruitment of DNA-PKcs induces inward translocation of Ku protein. Nucleic Acids Res 27(24):4679–4686
76. DeFazio LG, Stansel RM, Griffith JD, Chu G (2002) Synapsis of DNA ends by DNA-dependent protein kinase. EMBO J 21(12):3192–3200
77. Lieber MR, Lu H, Gu J, Schwarz K (2008) Flexibility in the order of action and in the enzymology of the nuclease, polymerases, and ligase of vertebrate non-homologous DNA end joining: relevance to cancer, aging, and the immune system. Cell Res 18(1):125–133
78. Jeggo P, O'Neill P (2002) The Greek goddess, Artemis, reveals the secrets of her cleavage. DNA Repair (Amst) 1(9):771–777
79. Chappell C, Hanakahi LA, Karimi-Busheri F, Weinfeld M, West SC (2002) Involvement of human polynucleotide kinase in double-strand break repair by non-homologous end joining. EMBO J 21(11):2827–2832
80. Perry JJ, Yannone SM, Holden LG et al (2006) WRN exonuclease structure and molecular mechanism imply an editing role in DNA end processing. Nat Struct Mol Biol 13(5):414–422
81. Bahmed K, Seth A, Nitiss KC, Nitiss JL (2011) End-processing during non-homologous end-joining: a role for exonuclease 1. Nucleic Acids Res 39(3):970–978
82. Ahnesorg P, Smith P, Jackson SP (2006) XLF interacts with the XRCC4-DNA ligase IV complex to promote DNA nonhomologous end-joining. Cell 124(2):301–313
83. Boland CR, Luciani MG, Gasche C, Goel A (2005) Infection, inflammation, and gastrointestinal cancer. Gut 54(9):1321–1331

Chapter 3
Resistance and DNA Repair Mechanisms of Cancer Stem Cells: Potential Molecular Targets for Therapy

Aamir Ahmad, Yiwei Li, Bin Bao and Fazlul H. Sarkar

Abstract Cancer stem cells (CSCs) are small subpopulations of cells within tumors that are intricately related to both *de novo* and acquired resistance to conventional therapies leading to tumor recurrence and metastasis. A majority of cancers initially respond to chemotherapeutic agents, as well as radiation therapy, but eventually develop resistance. An increased understanding of CSCs has led to the discovery that current treatments target the differentiated cancer cells leaving the CSCs unscathed due to their robust signaling pathways. Further, maintenance of genomic fidelity is important for normal functioning and survival of cells, including cancer cells and the CSCs. In this chapter, we will discuss several such pathways/phenomena which help CSCs resist therapies. These include increased quiescence and up-regulated drug transporters, activated DNA repair mechanisms and activation of several key cellular signaling pathways (Fig. 3.1). A better understanding of these resistance pathways is a necessary prerequisite towards the ultimate goal of developing novel strategies specifically targeting CSCs. Better designed therapies could ultimately reverse their resistance and thereby eliminate the potential of tumor recurrence and metastasis.

3.1 Introduction

It is estimated that in the United States alone a total of 1,596,670 new cancer cases will be reported in the current year and 571,950 patients will succumb to this disease [1]. In 2008, these numbers stood at more than 12 million cases and more than

A. Ahmad (✉) · Y. Li · B. Bao · F. H. Sarkar
Department of Pathology, Karmanos Cancer Institute, Wayne State University
School of Medicine, Detroit, MI, USA
e-mail: ahmada@karmanos.org

Y. Li
e-mail: yiweili@med.wayne.edu

B. Bao
e-mail: baob@karmanos.org

F. H. Sarkar
Department of Oncology, Karmanos Cancer Institute, Wayne State
University School of Medicine, Detroit, MI, USA
e-mail: fsarkar@med.wayne.edu

L. A. Mathews et al. (eds.), *DNA Repair of Cancer Stem Cells*,
DOI 10.1007/978-94-007-4590-2_3,

7 million deaths globally [2]. These statistics suggest that cancer remains the leading cause of deaths world-wide. Cancer is a disease characterized by uncontrolled cell growth which invariably is due to the manifestation of deregulated signaling pathways. The aberrations in cellular machinery result from genetic modifications that are increasingly accumulated. A number of signaling pathways have been implicated as the causes of cellular transformation of normal cells into tumor cells. It has also been suggested that a small population of cells, called cancer stem cells (CSCs), are critically important for the existence of tumors. These specialized cells are intricately linked with drug resistance, a characteristic of cancer cells that eventually leads to the development of tumor metastasis that is responsible for the demise of patients diagnosed with cancer.

3.2 CSCs and Relevant Cell Signaling Pathways

Stem cells are characterized by their ability to differentiate, giving rise to a variety of cell types. CSCs are found within populations of cancer cells or tumors and they possesses the capacity to self-renew and produce heterogeneous lineages of cancer cells [3]. CSCs, by virtue of being stem cells, have tumor-initiating capabilities. CSCs are now believed to persist in tumors as distinct populations that are fundamentally associated with drug resistance, tumor recurrence and metastasis. Recent evidence suggests that conventional therapies which target only the differentiated cells without affecting CSCs, leave dangerous cells capable of forming tumor masses. The role of Notch [4–6], Wnt [7] and sonic hedgehog (shh) [8, 9] signaling in cancer cells, and in particular CSCs, has been advocated for controlling survival signals in these cells [10, 11]. These signaling pathways offer attractive targets for the killing of CSCs and are being extensively investigated for their potential role in cancer therapy. Further, DNA repair mechanisms are also being studied for their role in the survival of CSCs [12, 13]. Additionally, these survival mechanisms also play an important role in the CSCs-modulated resistance to various therapies [14].

3.3 How Do CSCs Manage to Survive?

Cancer is a diverse disease with inherent heterogeneity. A number of therapeutic regimes are employed in clinics to manage cancer patients based on specific tumor type and subtype. While most of these therapies seem to be effective initially, a significantly large number of cancer patients eventually develop resistance against the drugs, also termed acquired chemo-resistance [15]. A number of theories have been put forward for the development of such acquired chemo-resistance, and the presence of CSC populations is one of them. There is also a growing acceptance that CSCs are the determinants of resistance to radiation therapy. It is believed that CSCs are resistant to drugs and radiation, and although most of the cancer cells are killed by the drugs, CSCs still survive [16]. Therefore, it is important to fully understand the mechanisms that could make CSCs refractory to conventional therapeutics. In the next sections we have outlined a few mechanisms believed to be responsible for the survival of CSCs.

3.3.1 *Quiescence*

Quiescence is a state of temporary inactivity. Since the chemotherapeutic regimes largely target rapidly proliferating cancer cells, it is believed that the ability of CSCs to proliferate slowly with intermittent phases of quiescence helps them evade the toxic effects of conventional anti-cancer therapeutics. It has, therefore, been suggested that targeting specific signaling pathways that mediate quiescence of CSCs might be an effective strategy for cancer therapy [17]. While there is some evidence that directly connects CSCs with the state of quiescence, as described below, the phenomenon of quiescence is characterized more extensively in normal stem cells [18, 19]. The connection of quiescence with CSCs stems from the fact that CSCs share many molecular similarities with normal stem cells [17] and as such, complicates the development of CSC-specific therapeutics.

In a pancreatic cancer model, there is direct evidence supporting the quiescence of CSCs. CSC subpopulations in BxPC-3 and Panc03.27 cells, marked either by the presence of surface antigens CD24 and CD44; or CD133 or by ALDH activity, represented slow cycling cells [20]. These slowly dividing cells were found to survive chemotherapeutic treatment, exhibited an increased invasive and tumorigenic potential, were able to recreate the initial heterogeneous tumor cell population and showed evidence of epithelial to mesenchymal transition (EMT) through an increase in the mesenchymal markers vimentin, snail, and twist. In an ovarian cancer model, it has been demonstrated [21] that the CD24(+) subpopulation of cells, representing the CSC phenotype, proliferates more slowly than the bulk tumor cells suggesting the quiescent nature of CSCs. As a proof-of-concept documenting that this subpopulation represented CSCs, CD24(+) cells were increasingly chemo-resistant, exhibited ability to self-renew and differentiate, and were found to be tumorigenic expressing higher levels of genes at the mRNA level that define "stemness" such as nestin, β-catenin, Bmi-1, Oct3/4, Notch-1 and Notch-4. Moreover, in a melanoma model [22], a very slow-cycling subpopulation with doubling time of more than 4 weeks was identified as well, which most likely represents CSCs. This subpopulation was characterized by the identification of H3K4 demethylase JARID1B as a biomarker. The knockdown of JARID1B was found to inhibit tumor growth and results in loss of the proliferation potential which supports the stemness of this subpopulation that expressed the biomarkers. This study supported the connection between CSCs and quiescence in melanoma cancer cells.

Thus, it appears that quiescence may represent a distinguishing feature of CSCs which provides them some degree of survival against the toxic effects of anti-cancer therapeutics that primarily target rapidly dividing cells. In addition to the pancreas, ovarian and melanoma cancer models, as discussed above, there is also emerging evidence suggesting a link between quiescence and CSCs in leukemia [23, 24] and intestinal [25, 26] cancers. As evidence continues to emerge, we will be able to gauge if there is a universal connection between quiescence and CSCs and, if so, there will be a need to develop appropriate model systems in order to study the processes of quiescence as potential targets for anti-cancer therapy. To that end, exposure to hypertonic medium has been suggested as a method to rapidly induce dormancy in prostate cancer cells [27]. With the development of experimental procedures that help simulate quiescence, the role of quiescence in sustenance of CSCs will be better understood, and such understanding will lead to more effective targeting of these pathways in overcoming therapeutic resistance.

3.3.2 Up-Regulation of ABC Transporters

ATP-binding cassette (ABC) transporters are transmembrane proteins which utilize the energy of ATP hydrolysis to transport various molecules across cellular membranes. They are named ABC transporters based on the sequence and organization of their ATP-binding cassette (ABC) domain(s), also known as nucleotide binding folds (NBFs) [28]. The role of ABC transporters in chemoresistance of cancer cells has long been advocated [29, 30] and there is ample literature available on the functionality of ABC transporters.

A number of mechanisms are now known which lead to resistance to anticancer drugs [29]. Some drug-resistance mechanisms interfere with the delivery of drugs to tumor cells while other drug-resistance mechanisms arise within the cancer cells leading to alterations in drug sensitivity. ABC transporters belong to the latter group. Drug resistance mediated by ABC transporters is not drug-specific and is believed to mediate resistance against entire classes of drugs. This is the reason why a cancer cell exposed to a specific drug might develop resistance not only against that particular drug, but also against other drugs which are structurally and/or mechanistically related. ABC transporters play important roles in resistance to natural-product hydrophobic drugs. For example, it has been shown that resistance against natural-product anticancer drugs paclitaxel, doxorubicin, or vinblastine is frequently due to increased expression of ABC transporters [29, 31]. Increased expression of ABC transporters increases the drug efflux from cancer cells leading to diminished intracellular concentrations of the chemotherapeutic drug. The lower concentrations of drug are not cytotoxic enough to effectively induce apoptosis or other cellular signaling damage that might otherwise result in cell death.

Early observations suggested that the ABC transporters ABCB1 and ABCG2 are specially up-regulated in hematopoietic stem cells, which provides strong evidence supporting the role of the ABC transporter family in conferring a stem cell-like phenotype [31, 30–34]. Thus, it is not surprising that these transporters are implicated in the drug resistance of CSCs [35, 36]. Since drug resistance of CSCs is frequently associated with ABC transporters, it can be argued that the inhibition of ABC transporters could be one of the approaches applied to render CSCs sensitive to chemotherapeutic drugs. Indeed, this was observed in squamous cell carcinoma [37] where blockage of ATP-dependent transporters led to chemo-sensitization of stem cell-like side populations of cells. Increased activity of ABC transporters in side populations of glioma has also been reported [38], and this study further showed the existence of Akt-dependent activation of ABCG2 and the loss of PTEN in CSC side population. Interestingly, treatment with temozolomide resulted in the enrichment of these subpopulations because of the up-regulation of ABC transporter ABCG2 in CSCs leading to resistance to therapy. It is important to note that studies looking at the side-population as a CSC marker are likely to find effects of the drug transporters since this is how the population is defined. Also, as suggested below, stem cell markers are known to induce the expression of ABC transporters, which might also explain the increased activity of these transporters in the side populations.

Studies of hepatocellular carcinoma proposed that the cells dual positive for CD133 and CD44 represent CSCs [39]. As expected, CD133(+)/CD44(+) cells were observed to be more resistant to chemotherapeutic agents and the up-regulation of ABC transporters ABCB1, ABCC1 and ABCG2 was believed to be the mechanism of drug resistance in these CSCs. Further evidence in support of ABC transporters in drug resistance of CSCs came from the observation that ABCB1-over-expressing side populations in ascites are enriched in patients whose ovarian cancer has relapsed following platinum-based chemotherapy compared to chemo-naive patients [40]. In leukemia, expression of the stem cell marker SALL4 has been shown to play an important role in determining the drug resistance of CSCs, and manipulations of SALL4 levels was correlated with response to chemotherapy [41]. Moreover, SALL4 has been shown to directly induce the ABC transporter ABCA3 via direct binding to its promoter as well indirectly inducing another ABC transporter ABCG2. The SALL4-induced induction of ABC transporters was also mechanistically linked with CSCs and their drug resistant phenotype. Although these observations functionally link stem cell markers with ABC transporters and suggest an induction of ABC transporters by stem cell markers, it is important to note that there is no evidence to suggest that ABC transporters are factors that define the 'stemness' of CSCs. In support of the direct evidence suggesting that ABC transporters do not play a role in the identity of CSCs, it has been observed that knockouts of genes for ABCG2, ABCB1, ABCC1 or combinations of these genes result in viable, fertile mice with normal stem cell populations [30, 42].

The role of ABC transporters in drug resistance of cancer cells has been widely accepted and more recent data suggests that these transporters are up-regulated in CSCs as well. ABC transporters work as guardians of CSCs by effectively effluxing the chemotherapeutic agent out of the cell [43]. Increased levels of ABC transporters in CSC populations equip them against the effects of chemotherapy, thus, ABC transporters are being pursued as valid targets for CSC therapies. It is anticipated that targeting ABC transporters in CSCs would be useful in overcoming therapeutic resistance, and thus would be useful for effective therapy [44].

3.3.3 Resistance to Irradiation

Radiation therapy remains the cutting-edge treatment option for multiple human malignancies. It is believed that the CSC subpopulations within the tumor are able to resist the damaging effects of radiation [45]. As a result, when the tumors are subjected to radiation therapy as part of the therapeutic regimen, the CSCs still thrive, contributing to tumor recurrence and metastasis [46]. Accumulating evidence suggests that a higher proportion of CSCs correlates with increased radio-resistance [47]. A majority of such studies have been carried out in cell lines using CSC markers, appropriate for a particular type of cancer, and an association between CSC population and resistance to radiation has been observed in glioma [48] as well as breast [49, 50] cancer models. Although in vivo validations of these observations are not readily available owing to the technical challenges associated with the designing of such studies; the available data strongly support an important role of CSCs in determining resistance to radiotherapy. We will summarize the biology of radiation resistance of CSCs in the following paragraphs.

In one of the earlier reports connecting CSCs with resistance to radiation therapy [48] it was shown that glioma CSCs exhibit increased DNA repair capacity which helps to reverse the DNA damage induced by radiation. Exposure to radiation resulted in enrichment of cells expressing the CSC marker CD133, and thus this study provided evidence linking CSCs with DNA damage repair and radio-resistance. Additionally, these CSCs also had higher expression of VEGF leading to increased angiogenesis [51] which could also explain the role of CSCs in response to therapy [52]. Therefore, DNA damage response and repair machinery represents an attractive target for enhancing the efficacy of CSCs targeted therapies [53]. In another report, the authors have suggested increased DNA repair machinery in CSCs lead to resistance to radiation whereby the DNA double strand break response machinery, particularly ATM, was reported to be activated in glioma CSCs [54]. ATM was also found to be induced in CSC subpopulations of breast cancer cells, MCF-7 and MDA-MB-231 [55]. These subpopulations demonstrated increased sphere forming capacity and were also resistant to radiation therapy. Interestingly, the non-homologous end joining DNA repair activity was not observed to be modulated in these subpopulations, suggesting a differential activation of DNA repair machinery in breast cancer CSCs.

One of the multiple signaling pathways activated in CSCs that determine their resistance to radiation is heat shock protein 27 (hsp27) that drives radiation resistance of breast CSCs through regulation of EMT and NF-κB signaling [56]. An earlier report documented similar findings linking EMT and radiation resistance with ovarian CSCs [57]. A cooperative modulation of gene regulation mediated by the transcriptional repressors snail and slug led to the induction of EMT, resistance to p53-mediated apoptosis and maintenance of self-renewal capacity. In rectal CSCs, expression of CD133, Oct4 and Sox2 were found to be elevated in tumors that are resistant to radiation therapy, which are associated with CSC markers and poor patient survival [58]. Finally, the CSC markers CD133, Sox2, Bmi and Nestin have been reported to be highly expressed in medulloblastoma CSCs that are increasingly resistant to radiotherapy, as well as TRAIL-induced apoptosis [58].

In a recent study comparing parental and radio-selected lung cancer cells A549 and breast cancer cells SKBR3 the authors have shown differential expression of Oct4 and Sox2, however the differences did not correlate with resistance to radiation [60]. The only CSC marker that was found to be correlated with radiation-resistance was ALDH1. These results highlight differences in individual CSC markers that might be relevant in a context-dependent manner, although the involvement of CSCs and radio-resistance is becoming clearer. Another report has also identified elevated ALDH levels in breast cancer CSCs and it was found to be responsible for radiation resistance [61]. The other factors that have been linked with radio-resistance of CSCs are nicotinamide N-methyltransferase expression [62], maternal embryonic leucine zipper kinases [63] and STAT3 signaling [64].

Glioma remains one of the highly investigated malignancies in the context of resistance to radiation therapy where CSCs are involved, as previously mentioned [48]. The CSC markers CD133, Nestin and Musashi were reported to be induced in gliomas exposed to radiation, which clearly suggested the functional significance of these markers in radio-resistance [65]. Furthermore, lymphocyte-specific protein tyrosine kinase (LCK) was determined to be important for this phenomenon because selective targeting of LCK resulted in the suppression of activation of not only these three CSC markers, but also other CSC markers such as Notch-2 and Sox2. The role of cyclooxygenase-2 (cox-2) in the enrichment of CD133 positive CSCs after radiation therapy has been reported where a cox-2 inhibitor, celecoxib, was found to significantly inhibit CD133 expressing cells leading to radio-sensitization of glioblastoma cells [66]. Another recent report has suggested defective autophagy to be a factor that contributes to radio-resistance of glioma CSCs [67]. It has been shown that the use of rapamycin could induce autophagy in glioma CSCs, thereby increasing their sensitivity to radiation therapy.

In addition to factors discussed above, a few other molecules/pathways have also been suggested to influence the resistance to radiotherapy in CSCs. For example, the Wnt signaling pathway has been associated with CSC phenotype, especially because activation of Wnt pathway was found in glioma cells isolated from mice after in vivo radiation [68]. These cells were also enriched for the CSC marker Sox2. This study provided direct evidence implicating the Wnt signaling pathway in mediating drug resistance of CSCs. In a recent study, the role of microRNAs (miRNAs) has been implicated in radiation resistance of CSCs where cells transfected with miR-145 were found to exhibit reduced tumorigenicity and stemness, leading to the reversal of radio-resistance [68]. This investigation was based on the realization that miR-145 is a suppressor of the CSC markers Oct4 and Sox2. We have discussed the role of miRNAs in CSC phenotype later in this chapter in relation to their regulation by natural agents.

Deregulated expression of many genes, including developmentally regulated genes, is the "hallmark" of CSCs. In addition, these cells are highly resistant to conventional therapeutics, and thus molecules that are involved in cell cycle regulation, DNA damage and repair are considered important targets for overcoming such resistance [68]. Therefore, it has been proposed that targeting CSCs might be an effective strategy to overcome resistance to radiation [68–68] and other therapeutics as summarized above.

3.3.4 DNA Damage-Repair in CSCs

Summarized in Chap. 2, it is clear that a number of factors are believed to cause DNA damage including chemicals, UV radiation, viruses and reactive oxygen species (ROS). DNA can be 'damaged' in many different ways including the loss of purine or pyrimidine bases leading to formation of 'abasic sites', the bases may lose their amino groups, bases may be oxidized, bases may be alkylated or there might be actual strand breaks in the DNA. Cells with irreparable DNA damage either enter senescence or they are eliminated via induction of apoptosis [77]. DNA damage involves an actual change in the normal structure of DNA and these changes are detected by the cellular DNA repair machinery. In the case of DNA damage by oxidation, alkylation or hydrolysis of bases, a 'base excision repair' mechanism repairs the damage by removing the damaged bases and replacing it with normal bases through a series of steps catalyzed by specific enzymes. In case of damage to one DNA strand, the second complimentary strand serves as the template for the synthesis of normal 'repaired' strand. In addition to taking care of damaged DNA that involves just a single strand, there are additional DNA repair mechanisms in place which repair double strand breaks [78, 79]. In order to ensure a very efficient repair of DNA damage, CSCs have multiple DNA repair mechanisms in place involving a multitude of factors/proteins [12]. Cancer cells evade cell cycle check points leading to uncontrolled DNA replication and cell growth, therefore, it is not surprising that in addition to CSCs, cancer cells are characterized by their capacity for highly efficient DNA repair mechanism [80, 81]. Consequently, DNA repair capacity is a subject of clinical assessment to determine cancer development and response to therapy [82].

It is interesting to note that the well-organized DNA repair machinery which plays a key role in ensuring genomic stability of embryonic stem cells (ESCs), thereby being a 'life-saver', actually has a dark side when studied in the context of CSCs [83]. The very cellular machinery which makes sure that the correct information is relayed to differentiated cells and cell lineages in ESCs turns evil in CSCs wherein it works to reverse the DNA damage induced by therapeutic regimes. The role of CSCs in metastases as well as in drug resistance of human cancers is becoming more and more apparent, and the importance of DNA repair mechanisms in CSCs is being recognized.

3.3.4.1 DNA Repair and Drug Resistance in Cancer and CSCs

A number of anti-cancer drugs used in the clinic for the treatment of cancer patients are known to target genomic stability. This approach stems from the realization that cancer cells have an increased rate of DNA replication. Targeting genomic stability should therefore be an effective strategy to drive cancer cells into the path of apoptotic cell death. For example, cisplatin is a very common and effective chemotherapeutic agent, wherein platinum binds to the purines in DNA structure and distorts the double helix leading to DNA damage. In fact, not only cisplatin, but other 'platinum' drugs such as oxaliplatin and carboplatin are capable of inducing DNA damage as well [84]. The cancer cells that have compromised DNA repair mechanism are destroyed through apoptotic-induction. However, it is also known that within 6–8 hours, the DNA damage repair mechanisms identify the cisplatin-induced damage to DNA and approximately 2–3 of the platinum-DNA adducts are removed within 21 hours [87], which is believed to be responsible for treatment failure. Moreover, it has been suggested that several independent DNA repair mechanisms can restore the integrity of bases that are alkylated by alkylating anti-cancer agents, such as in the treatment of glioblastoma [86]. Such efficient repair of DNA damage contributes to drug resistance and subsequent treatment failure.

3.3.4.2 Enhanced DNA Repair in CSCs

It has been postulated that CSCs have more efficient DNA repair mechanisms in place to ensure genomic stability compared to cancer cells. To effectively target CSCs, it is prudent to target their DNA repair mechanisms. In a study suggestive of such action of HDAC inhibitors [85], it was shown that HDAC inhibitors SAHA and MS-275 had a profound effect on several genes that are involved in various DNA repair mechanisms in mesenchymal stem cells. As a consequence, the markers of DNA damage accumulated when these cells were treated with HDAC inhibitors. It is expected that with increased DNA damage, through impaired repair mechanisms, CSCs can be made sensitive to anticancer drugs that function through DNA damage.

In a study that was designed to directly evaluate DNA repair in cancer vs. CSCs, a number of stem and non-stem glioma cell lines were directly compared for their relative DNA repair potential [73, 81]. Interestingly, none of the DNA repair pathways tested; DNA base excision or single strand break repair, were observed to be elevated in glioma CSCs, compared to glioma cells. Based on the observations regarding doubling time and cell cycle regulatory proteins, it was proposed that an increased expression of cell cycle checkpoint kinases CHK1 and CHK2 leads to a delayed cell cycle in glioma CSCs. Such delay in cell cycle, in turn, results in a significantly increased time that is available for DNA repair in CSCs [12]. Activation of CHK1 has recently been demonstrated in non-small-cell lung cancer CSCs as well [88]. In a report that further supports this hypothesis, a similar mechanism was shown to operate in leukemia CSCs [89]. In this model, the role of cell cycle inhibitor p21 was proposed, suggesting the existence of an effective DNA repair mechanisms in the CSCs. The p21-mediated inhibition of cell cycle was found to provide sufficient time for CSCs to repair accumulated DNA damage. These studies point to an indirect mechanism whereby the CSCs buy more time for an efficient DNA repair through delayed cell cycle progression.

Although these initial studies did not find elevated DNA repair mechanisms in CSCs, a later study [54] reported elevated DNA double strand break response machinery in glioma CSCs, thus providing direct evidence to suggest increased DNA repair activity in CSCs. In this study, a polycomb group protein was reported to be enriched in glioma CSCs where it co-purified with multiple factors from DNA double strand break response and non-homologous end joining repair mechanisms. Its deficiency led to severely impaired DNA repair in CSCs rendering them sensitive to radiation. In another report on the elevated DNA repair mechanisms in CSCs [90], the role of a DNA-dependent protein kinase catalytic subunit (DNA-PKcs) was suggested during DNA double strand repair process. In breast cancer CSCs derived from MCF-7, MDA-MB-231 cell lines, as well as primary cultures of patient breast cancer cells, activation of the DNA double strand break response via ATM was reported in CSCs with no significant changes observed in non-homologous end joining repair response machinery [55]. The CSCs in this study were identified on the basis of CD44(+)/CD24(− or low) expression. Additionally, the DNA single strand break repair pathway was found to be significantly active in another investigation that looked at CD44(+)/CD24(−/low) mammospheres derived from MCF-7 cells [91]. Interestingly, this study did not find any significant changes in DNA double strand break repair machinery. Finally, in an earlier study involving breast CSCs, the CSC subpopulation [Lin(-)CD29(H)CD24(H)] was reported to express up-regulated levels of genes involved in DNA damage response and repair [92].

Experimentally induced pluripotent cells and embryonic stem cells have significantly enhanced levels of DNA repair mechanisms [93]. Emerging evidence indicates that the CSCs are also characterized by such elevated levels of efficient DNA repair processes [12]. As summarized above, studies in several different cancer models generally support this notion, although the evidence seems to be contradictory at times. One observation that stands out is that CSCs have markedly reduced DNA damage, which is either due to the direct activation of DNA repair processes or through delayed cell cycle which allows extra time for DNA repair. Even within the DNA repair processes, different mechanisms are being validated to be functional in different model systems, as well as within the same model. It is quite possible that the efficient DNA repair in CSCs is a result of several such processes that operate in harmony. Inclusion of inhibition of DNA repair mechanisms as a component in combinational therapy is therefore likely to target CSCs with increased efficacy.

3.4 Targeting CSCs with Natural Compounds

Compounds that are found in plants and other natural resources are well-established as anticancer agents [92–95]. Since these compounds are part of a normal diet, they present much reduced systemic toxicity when compared to synthetic compounds. With the acceptance of an important role of CSCs in the sustenance of human cancers, emerging data points towards the ability of natural compounds for targeting CSCs [98, 99].

One such compound, the naturally occurring phytochemical, parthenolide (PTL) has been documented to target CSCs in multiple human cancer models [100]. In an earlier report, PTL was observed to preferentially target acute myelogenous leukemia progenitor and stem cell populations in a leukemia model through inhibition of NF-κB, activation of p53 and induction of reactive oxygen species (ROS) [101]. Other reports have further verified such CSC-targeting action of PTL in leukemia models [102, 103]. Involvement of NF-κB signaling in mediating the biological effects of PTL against breast CSCs has also been reported [104]. In addition, PTL has been found to inhibit the growth of CSC side populations derived from breast cancer cell lines, MCF-7 and MDA-MB-231 [105]. Subsequent to these findings, PTL has been shown to be cytotoxic against prostate CSCs that were isolated from prostate cancer cell lines DU145, PC3, VCAP and LAPC4, as well as from the primary prostate tumor initiating cells [106]. The non-receptor tyrosine kinase src, and several src signaling components such as Csk, FAK, β1-arrestin, FGFR2, PKC, MEK/MAPK, CaMK, ELK-1 and ELK-1-dependent genes were found to be the targets of PTL against prostate CSCs. More recently, PTL has been shown to target CSCs in osteosarcoma [107] and myeloma [108] models.

In addition to studies on the effect of PTL on prostate derived CSCs [108], the Farrar group has reported the ability of another natural compound, gossypol to target prostate CSCs [109]. Gossypol, the phytochemical produced by cotton plants, effectively inhibited prostate tumor-initiating cell-driven tumor growth in a non-obese diabetic/severe combined immune-deficient xenograft model through induction of DNA damage and activation of p53. As discussed above, the DNA repair machinery in CSCs is particularly efficient and its effective targeting is one of the mechanisms through which any putative agent could target CSCs, and as such could become a useful strategy for the killing of CSCs and thereby eliminate tumors.

Another natural compound that has been shown to target CSCs in the last couple of years is the anticancer agent resveratrol. Chemically, resveratrol is 3,5,4′-trihydroxy-trans-stilbene, a naturally occurring polyphenol and a phytoalexin produced by several plants. It is typically found in the skin of red grapes and is present in red wines. In normal stem cells, there is evidence to support the role of resveratrol in accelerating DNA repair mechanisms thereby helping stem cells survive [110]. The initial report on CSCs-targeting activity of resveratrol suggested its ability to inhibit medulloblastoma CSCs-associated proliferation and tumorigenicity [111]. Later, resveratrol was reported to inhibit glioma CSCs-induced tumorigenicity through its inhibitory action on STAT3 signaling, which led to increased radio-sensitivity [112]. Finally, resveratrol has been observed to inhibit mammosphere formation by CD24(–)/CD44(+)ESA(+) CSCs isolated from estrogen receptor-positive as well as estrogen receptor-negative breast cancer cells [113]. In human pancreatic CSCs, characterized by CD133(+)/CD44(+)/CD24(+)/ESA(+), resveratrol has been shown to inhibit self-renewal capacity [114]. Resveratrol also inhibited Nanog, Sox-2, c-myc and Oct-4, the markers of 'stemness', as well as ABC transporter, ABCG2 in CSCs. Furthermore, resveratrol inhibited CSC's migration and invasion and attenuated the markers of EMT. The inhibition of stem cell markers is direct evidence in support of the biological action of resveratrol against CSCs and additionally, down-regulation of ABC transporter further supports the anti-CSC activity of resveratrol because ABC transporters, as discussed above, are frequently up-regulated in CSCs.

Sulforaphane, found in cruciferous vegetables, is another natural compound that has been investigated for its action against CSCs [99]. In breast cancer CSCs, sulforaphane has been found to be very effective in inhibiting ALDH-positive CSC population that is mediated through down-regulation of wnt/β-catenin signaling pathway [115]. Sulforaphane has also been shown to inhibit ALDH activity of pancreatic CSCs [116]. This study identified NF-κB signaling as the major target of sulforaphane leading to the potentiating effect of this natural compound on the activity of kinase inhibitor sorafenib. In another report, the authors have documented potentiating effects of sulforaphane leading to effective inhibition of CSCs [117] where it was shown to synergistically induce the cytotoxic effect of cisplatin, gemcitabine, doxorubicin and 5-flurouracil against CSCs of prostate and pancreatic origins. The combinations were observed to be significant in inhibiting the ALDH1 activity and also Notch-1 and c-Rel expression, indicators of "stemness" of CSCs. In yet another study, the synergistic effect of sulforaphane has been reported in combination with another natural agent, quercetin, in inhibiting pancreatic CSCs [118]. An involvement of stem cell marker Nanog was suggested based on the observation that silencing of Nanog further enhanced the inhibitory action of these natural compounds on the self-renewing capability of CSCs mediated through the targeting of EMT. The natural compound quercetin, a flavonoid obtained from many dietary fruits and vegetables, has been shown to be active by itself against pancreatic CSCs [119]. In addition to its synergistic effects in combination with sulforaphane, as discussed above, quercetin has also been reported to show synergistic activity when combined with green tea polyphenol, epigallocatechin-3-gallate (EGCG), which led to cytotoxic activity against prostate [120] and pancreatic CSCs [121].

The one natural compound that has been extensively studied in relation to its ability to affect CSCs is the turmeric-derived compound, curcumin. A number of investigations, including studies from our own laboratory, have described such action of curcumin as well as its novel synthetic derivative. In one of the earliest reports on this topic, Wicha and co-workers demonstrated the ability of curcumin to inhibit ALDH activity, mammosphere formation and inhibition of Wnt signaling in breast CSCs [122]. Curcumin was also reported to be effective against colon CSCs as evidenced by decreased CD44 and CD166 positive colon cancer HCT-116 and HT-29 cells [123]. Further, it inhibited the side populations of glioma cells, suggesting its activity against glioma CSCs [124]. While such action of curcumin is promising, curcumin suffers from limited bioavailability *in vivo*, and thus the use of curcumin in human patients has been disappointing. To overcome this limitation, we have synthesized a derivative of curcumin, the difluorinated curcumin, CDF [125] which showed improved bioavailability and tissue distribution [126]. This novel synthetic derivative has actually demonstrated increased activity against colon CSCs than its precursor curcumin, whereby we observed reduction of CD44 and CD166 in chemo-resistant colon cancer cells [127]. We also showed, for the first time, the effects of CDF against pancreatic CSCs [128], whereby we have shown that this compound can significantly inhibit pancreatospheres and reduce CSC markers CD44 and EpCAM in gemcitabine resistant pancreatic cancer cells MIAPaCa-2 with high proportions of CSCs. Further, in another recent report, we have demonstrated the ability of CDF to inhibit gliomas methyltransferase EZH2, a determinant of stem cells survival and function, leading to an effective inhibition of pancreatic CSCs [129]. Mechanistically we found that re-expression of the miRNA miR-101 was sufficient in limiting EZH2 expression, and that the administration of CDF inhibited tumor growth *in vivo* through reduced expression of EZH2, Notch-1, CD44, EpCAM and Nanog (all determinants of "stemness"), which was mediated by increased expression of let-7, miR-26a and miR-101. Our study, on one hand, demonstrated the ability of CDF to effectively target CSCs, and on the other hand, also underscores the importance of miRNAs in regulating CSCs integrity and function. Consistent with our findings, other investigators have also shown the potential of curcumin formulations [130] as well as curcumin analogue [131] in the inhibition of CSCs of gliomas and colon origin, respectively.

In summary, the data from our laboratory and others clearly suggests that natural compounds could be very effective in targeting CSCs by attenuating multiple targets, especially by re-expression of miRNAs that are typically lost in cancer and in CSCs. The relative non-toxic nature of these compounds is very attractive for further clinical investigation in the targeted elimination of CSCs to eradicate tumors.

3.5 Conclusions and Perspectives

CSCs are hard to kill by conventional therapeutics, although they account for less than 1 % of total cell population in a tumor, which is believed to be the cause of tumor recurrence and metastasis. In this chapter we have discussed some of the mechanisms that help CSCs survive cancer therapies and these include quiescence, up-regulated ABC transporters, efficient DNA repair machinery and several up-regulated cellular signaling pathways as succinctly illustrated in Fig. 3.1. Moreover, CSCs are also enriched after conventional therapy, suggesting that better understanding of resistance mechanisms of CSCs would be useful for further development of novel and targeted therapies for the killing of CSCs. As discussed here, significant progress has been made in the last few years; however, the challenges are mounting because of the complexity of multiple signaling pathways in the self-renewal, survival and proliferation of CSCs. Interestingly, natural agents are multi-targeting and therefore it is highly promising that natural agents could become an innovative approach for targeted killing of CSCs directly or by attenuating cellular signaling in order to make CSCs sensitive to conventional therapeutics. Among many signaling molecules, the roles of miRNAs are an emerging area of intense research especially because natural agents could be useful for the re-expression of miRNAs that are typically lost in cancers, and especially in CSCs. Therefore, targeted re-expression of miRNAs would likely be a novel approach for targeting "stemness" markers for the elimination of CSCs. The evidence so far is indicative of an important role of multiple mechanisms that define resistance of CSCs. It is essential to fully elucidate the intricate cross-talk

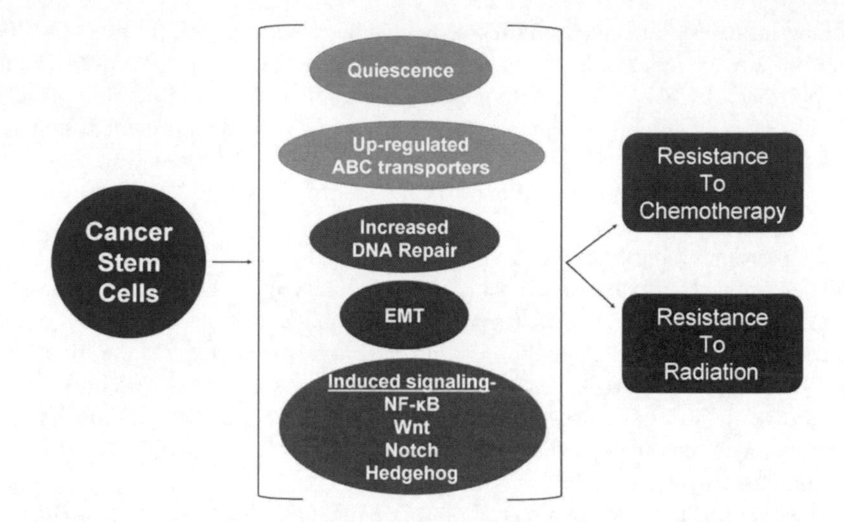

Fig. 3.1 Mechanisms that determine resistance of cancer stem cells to chemotherapy and radiation therapy

and molecular regulation of these mechanisms to develop effective treatment strategies. Further in-depth molecular pre-clinical and clinical investigations are warranted in order to exploit the role of novel therapeutic agents against resistance pathways in CSCs, which would likely be useful for eradication of human malignancies in the near future.

References

1. Siegel R, Ward E, Brawley O, Jemal A (2011) Cancer statistics, 2011: the impact of eliminating socioeconomic and racial disparities on premature cancer deaths. CA Cancer J Clin 61:212–236
2. Jemal A, Bray F, Center MM, Ferlay J, Ward E, Forman D (2011) Global cancer statistics. CA Cancer J Clin 61:69–90
3. Clarke MF, Dick JE, Dirks PB, Eaves CJ, Jamieson CH, Jones DL, Visvader J, Weissman IL, Wahl GM (2006) Cancer stem cells—perspectives on current status and future directions: AACR workshop on cancer stem cells. Cancer Res 66:9339–9344
4. Farnie G, Clarke RB (2007) Mammary stem cells and breast cancer—role of Notch signalling. Stem Cell Rev 3:169–175
5. Wang Z, Li Y, Banerjee S, Sarkar FH (2009) Emerging role of Notch in stem cells and cancer. Cancer Lett 279:8–12
6. Pannuti A, Foreman K, Rizzo P, Osipo C, Golde T, Osborne B, Miele L (2010) Targeting Notch to target cancer stem cells. Clin Cancer Res 16:3141–152
7. Katoh M, Katoh M (2007) WNT signaling pathway and stem cell signaling network. Clin Cancer Res 13:4042–4045
8. Katoh Y, Katoh M (2006) Hedgehog signaling pathway and gastrointestinal stem cell signaling network (review). Int J Mol Med 18:1019–1023
9. Medina V, Calvo MB, az-Prado S, Espada, J (2009) Hedgehog signalling as a target in cancer stem cells. Clin Transl Oncol 11:199–207

10. Cerdan C, Bhatia M (2010) Novel roles for Notch, Wnt and Hedgehog in hematopoesis derived from human pluripotent stem cells. Int J Dev Biol 54:955–963
11. Takebe N, Harris PJ, Warren RQ, Ivy SP (2011) Targeting cancer stem cells by inhibiting Wnt, Notch, and Hedgehog pathways. Nat Rev Clin Oncol 8:97–106
12. Mathews LA, Cabarcas SM, Farrar WL (2011) DNA repair: the culprit for tumor-initiating cell survival? Cancer Metastasis Rev 30:185–197
13. Blanpain C, Mohrin M, Sotiropoulou PA, Passegue E (2011) DNA-damage response in tissue-specific and cancer stem cells. Cell Stem Cell 8:16–29
14. Lacerda L, Pusztai L, Woodward WA (2010) The role of tumor initiating cells in drug resistance of breast cancer: implications for future therapeutic approaches. Drug Resist Updat 13:99–108
15. Wang Z, Li Y, Ahmad A, Azmi AS, Kong D, Banerje S, Sarkar FH (2010) Targeting miRNAs involved in cancer stem cell and EMT regulation: an emerging concept in overcoming drug resistance. Drug Resist Updat 13:109–118
16. Dean M, Fojo T, Bates S (2005) Tumour stem cells and drug resistance. Nat Rev Cancer 5:275–284
17. Moore N, Lyle S (2011) Quiescent, slow-cycling stem cell populations in cancer: a review of the evidence and discussion of significance. J Oncol (PMID: 20936110; doi: 10.1155/2011/396076, http://www.hindawi.com/journals/jo/2011/396076/)
18. Viale A, Pelicci PG (2009) Awaking stem cells from dormancy: growing old and fighting cancer. EMBO Mol Med 1:88–91
19. Li L, Bhatia R (2011) Stem cell quiescence. Clin Cancer Res 17:4936–4941
20. Dembinski JL, Kraus S (2009) Characterization and functional analysis of a slow cycling stem cell-like subpopulation in pancreas adenocarcinoma. Clin Exp Metastasis 26:611–623
21. Gao MQ, Choi YP, Kang S, Youn JH, Cho NH (2010) $CD24^+$ cells from hierarchically organized ovarian cancer are enriched in cancer stem cells. Oncogene 29:2672–2680
22. Roesch A, Fukunaga-Kalabis M, Schmidt EC, Zabierowski SE, Brafford PA, Vultur A, Basu D, Gimotty P, Vogt T, Herlyn M (2010) A temporarily distinct subpopulation of slow-cycling melanoma cells is required for continuous tumor growth. Cell 141:583–594
23. Forsberg EC, Passegue E, Prohaska SS, Wagers AJ, Koeva M, Stuart JM, Weissman IL (2010) Molecular signatures of quiescent, mobilized and leukemia-initiating hematopoietic stem cells. PLoS One 5:e8785
24. Ichihara E, Kaneda K, Saito Y, Yamakawa N, Morishita K (2011) Angiopoietin1 contributes to the maintenance of cell quiescence in EVI1(high) leukemia cells. Biochem Biophys Res Commun 416:239–245
25. Roth S, Fodde R (2011) Quiescent stem cells in intestinal homeostasis and cancer. Cell Commun Adhes 18:33–44
26. Buczacki S, Davies RJ, Winton DJ (2011) Stem cells, quiescence and rectal carcinoma: an unexplored relationship and potential therapeutic target. Br J Cancer 105:1253–1259
27. Havard M, Dautry F, Tchenio T (2011) A dormant state modulated by osmotic pressure controls clonogenicity of prostate cancer cells. J Biol Chem 286:44177–44186
28. Dean M, Allikmets R (1995) Evolution of ATP-binding cassette transporter genes. Curr Opin Genet Dev 5:779–785
29. Gottesman MM, Fojo T, Bates SE (2002) Multidrug resistance in cancer: role of ATP-dependent transporters. Nat Rev Cancer 2:48–58
30. Dean M (2009) ABC transporters, drug resistance, and cancer stem cells. J Mammary Gland Biol Neoplasia 14:3–9
31. Ueda K, Cardarelli C, Gottesman MM, Pastan I (1987) Expression of a full-length cDNA for the human "MDR1" gene confers resistance to colchicine, doxorubicin, and vinblastine. Proc Natl Acad Sci USA 84:3004–3008
32. Scharenberg CW, Harkey MA, Torok-Storb B (2002) The ABCG2 transporter is an efficient Hoechst 33342 efflux pump and is preferentially expressed by immature human hematopoietic progenitors. Blood 99:507–512
33. Kim M, Turnquist H, Jackson J, Sgagias M, Yan Y, Gong M, Dean M, Sharp JG, Cowan K. (2002) The multidrug resistance transporter ABCG2 (breast cancer resistance protein 1)

effluxes Hoechst 33342 and is overexpressed in hematopoietic stem cells. Clin Cancer Res 8:22–28
34. Lou H, Dean M (2007) Targeted therapy for cancer stem cells: the patched pathway and ABC transporters. Oncogene 26:1357–1360
35. An Y, Ongkeko WM (2009) ABCG2: the key to chemoresistance in cancer stem cells? Expert Opin Drug Metab Toxicol 5:1529–1542
36. Elliot A, Adams J, Al-Haj M (2010) The ABCs of cancer stem cell drug resistance. I Drugs 13:632–635
37. Loebinger MR, Giangreco A, Groot KR, Prichard L, Allen K, Simpson C, Bazley L, Navani N, Tibrewal S, Davies D, Janes SM (2008) Squamous cell cancers contain a side population of stem-like cells that are made chemosensitive by ABC transporter blockade. Br J Cancer 98:380–387
38. Bleau AM, Hambardzumyan D, Ozawa T, Fomchenko EI, Huse JT, Brennan CW, Holland EC (2009) PTEN/PI3K/Akt pathway regulates the side population phenotype and ABCG2 activity in glioma tumor stem-like cells. Cell Stem Cell 4:226–235
39. Zhu Z, Hao X, Yan M, Yao M, Ge C, Gu J, Li J (2010) Cancer stem/progenitor cells are highly enriched in CD133^{+} CD44^{+} population in hepatocellular carcinoma. Int J Cancer 126:2067–2078
40. Rizzo S, Hersey JM, Mellor P, Dai W, Santos-Silva A, Liber D, Luk L, Titley I, Carden CP, Box G, Hudson DL, Kaye SB, Brown R (2011) Ovarian cancer stem cell-like side populations are enriched following chemotherapy and overexpress EZH2. Mol Cancer Ther 10:325–335
41. Jeong HW, Cui W, Yang, Y, Lu J, He J, Li A, Song D, Guo Y, Liu BH, Chai L (2011) SALL4, a stem cell factor, affects the side population by regulation of the ATP-binding cassette drug transport genes. PLoS One 6:e18372
42. Schinkel AH, Smit JJ, van TO, Beijnen JH, Wagenaar E, van DL, Mol CA, van dV Robanus-Maandag EC, te Riele HP (1994) Disruption of the mouse mdr1a P-glycoprotein gene leads to a deficiency in the blood-brain barrier and to increased sensitivity to drugs. Cell 77:491–502
43. Moitra K, Lou H, Dean M (2011) Multidrug efflux pumps and cancer stem cells: insights into multidrug resistance and therapeutic development. Clin Pharmacol Ther 89:491–502
44. Lee CH (2010) Reversing agents for ATP-binding cassette drug transporters. Methods Mol Biol 596:325–340
45. Hambardzumyan D, Squatrito M, Holland EC (2006) Radiation resistance and stem-like cells in brain tumors. Cancer Cell 10:454–456
46. Vlashi E, McBride WH, Pajonk F (2009) Radiation responses of cancer stem cells. J Cell Biochem 108:339–342
47. Bauman M, Krause M, Hil R (2008) Exploring the role of cancer stem cells in radioresistance. Nat Rev Cancer 8:545–554
48. Bao S, Wu Q, McLendon RE, Hao Y, Shi Q, Hjelmeland AB, Dewhirst MW, Bigner DD, Rich JN (2006) Glioma stem cells promote radioresistance by preferential activation of the DNA damage response. Nature 444:756–760
49. Phillips TM, McBride WH, Pajonk F (2006) The response of CD24(−/low)/CD44^{+} breast cancer-initiating cells to radiation. J Natl Cancer Inst 98:1777–1785
50. Woodward WA, Chen MS, Behbod F, Alfaro MP, Buchholz TA, Rosen JM (2007) WNT/beta-catenin mediates radiation resistance of mouse mammary progenitor cells. Proc Natl Acad Sci USA 104:618–623
51. Bao S, Wu Q, Sathornsumete S, Hao Y, Li Z, Hjelmeland AB, Shi Q, McLendon RE, Bigner DD, Rich JN (2006) Stem cell-like glioma cells promote tumor angiogenesis through vascular endothelial growth factor. Cancer Res 66:7843–7848
52. Rich JN (2007) Cancer stem cells in radiation resistance. Cancer Res 67:8980–8984
53. Gieni RS, Ismail IH, Campbel S, Hendzel MJ (2011) Polycomb group proteins in the DNA damage response: a link between radiation resistance and "stemness". Cell Cycle 10:883–894
54. Facchino S, Abdouh M, Chato W, Bernier G (2010) BMI1 confers radioresistance to normal and cancerous neural stem cells through recruitment of the DNA damage response machinery. J Neurosci 30:10096–10111

55. Yin H, Glas J (2011) The phenotypic radiation resistance of CD44^{+}/CD24(−or low) breast cancer cells is mediated through the enhanced activation of ATM signaling. PLoS One 6:e24080
56. Wei L, Liu TT, Wang HH, Hong HM, Yu AL, Feng HP, Chang WW (2011) Hsp27 participates in the maintenance of breast cancer stem cells through regulation of epithelial-mesenchymal transition and nuclear factor-kappa B. Breast Cancer Res 13:R101
57. Kurrey NK, Jalgaonkar SP, Joglekar AV, Ghanate AD, Chaskar PD, Doiphode RY, Bapat SA (2009) Snail and slug mediate radioresistance and chemoresistance by antagonizing p53-mediated apoptosis and acquiring a stem-like phenotype in ovarian cancer cells. Stem Cells 27:2059–2068
58. Saigusa S, Tanaka K, Toiyama Y, Yokoe T, Okugawa Y, Ioue Y, Miki C, Kusunoki M (2009) Correlation of CD133, OCT4, and SOX2 in rectal cancer and their association with distant recurrence after chemoradiotherapy. Ann Surg Oncol 16:3488–3498
59. Yu CC, Chiou GY, Le YY, Chang YL, Huang PI, Cheng YW, Tai LK, Ku HH, Chiou SH, Wong TT (2010) Medulloblastoma-derived tumor stem-like cells acquired resistance to TRAIL-induced apoptosis and radiosensitivity. Childs Nerv Syst 26:897–904
60. Mihatsch J, Toulany M, Bareis PM, Grim S, Lengerke C, Kehlbach R, Rodeman HP (2011) Selection of radioresistant tumor cells and presence of ALDH1 activity in vitro. Radiother Oncol 99:300–306
61. Croker AK, Allan AL (2012) Inhibition of aldehyde dehydrogenase (ALDH) activity reduces chemotherapy and radiation resistance of stem-like ALDH(hi)CD44 (+) human breast cancer cells. Breast Cancer Res Treat 133:75–87
62. D'Andrea FP, Safwat A, Kassem M, Gautier L, Overgaard J, Horsman MR (2011) Cancer stem cell overexpression of nicotinamide N-methyltransferase enhances cellular radiation resistance. Radiother Oncol 99:373–378
63. Choi S, Ku JL (2011) Resistance of colorectal cancer cells to radiation and 5-FU is associated with MELK expression. Biochem Biophys Res Commun 412:207–213
64. Lin L, Fuchs J, Li C, Olson V, Bekaii-Saab T, Lin J (2011) STAT3 signaling pathway is necessary for cell survival and tumorsphere forming capacity in ALDH(+)/CD133(+) stem cell-like human colon cancer cells. Biochem Biophys Res Commun 416:246–251
65. Kim RK, Yoon CH, Hyun KH, Le H, An S, Park MJ, Kim MJ, Le SJ (2010) Role of lymphocyte-specific protein tyrosine kinase (LCK) in the expansion of glioma-initiating cells by fractionated radiation. Biochem Biophys Res Commun 402:631–636
66. Ma HI, Chiou SH, Hueng DY, Tai LK, Huang PI, Kao CL, Chen YW, Sytwu HK (2011) Celecoxib and radioresistant glioblastoma-derived CD133^{+} cells: improvement in radiotherapeutic effects. Lab Inves J Neurosurg 114:651–662.
67. Zhuang W, Li B, Long L, Chen L, Huang Q, Liang Z (2011) Induction of autophagy promotes differentiation of glioma-initiating cells and their radiosensitivity. Int J Cancer 129:2720–2731
68. Kim Y, Kim KH, Le J, Le YA, Kim M, Le SJ, Park K, Yang H, Jin J, Jo KM, Le J, Nam DH (2012) Wnt activation is implicated in glioblastoma radioresistance. Lab Invest 92:466–473
69. Yang YP, Chien Y, Chiou GY, Cherng JY, Wang ML, Lo WL, Chang YL, Huang PI, Chen YW, Shih YH, Chen MT, Chiou SH (2012) Inhibition of cancer stem cell-like properties and reduced chemoradioresistance of glioblastoma using microRNA145 with cationic polyurethane-short branch PEI. Biomaterials 33:1462–1476
70. Pajonk F, Vlashi E, McBride WH (2010) Radiation resistance of cancer stem cells: the 4 R's of radiobiology revisited. Stem Cells 28:639–648
71. Debeb BG, Xu W, Woodward WA (2009) Radiation resistance of breast cancer stem cells: understanding the clinical framework. J Mammary Gland Biol Neoplasia 14:11–17
72. Cripe TP, Wang PY, Marcato P, Mahller YY, Le PW (2009) Targeting cancer-initiating cells with oncolytic viruses. Mol Ther 17:1677–1682
73. Frosina G (2009) DNA repair and resistance of gliomas to chemotherapy and radiotherapy. Mol Cancer Res 7:989–999
74. Morrison R, Schleicher SM, Sun Y, Nierman KJ, Kim S, Sprat DE, Chung CH, Lu B (2011) Targeting the mechanisms of resistance to chemotherapy and radiotherapy with the cancer stem cell hypothesis. J Oncol 2011:941876

75. Roos WP, Kaina B (2006) DNA damage-induced cell death by apoptosis. Trends Mol Med 12:440–450
76. Xu Y, Price BD (2011) Chromatin dynamics and the repair of DNA double strand breaks. Cell Cycle 10:261–267
77. Rossetto D, Truman AW, Kron SJ, Cote J (2010) Epigenetic modifications in double-strand break DNA damage signaling and repair. Clin Cancer Res 16:4543–4552
78. Frosina G (2009) DNA repair in normal and cancer stem cells, with special reference to the central nervous system. Curr Med Chem 16:854–866
79. Ropolo M, Daga A, Griffero F, Foresta M, Casartelli G, Zunino A, Poggi A, Cappelli E, Zona G, Spaziante R, Corte G, Frosina G (2009) Comparative analysis of DNA repair in stem and nonstem glioma cell cultures. Mol Cancer Res 7:383–392
80. Jalal S, Earley JN, Turchi JJ (2011) DNA repair: from genome maintenance to biomarker and therapeutic target. Clin Cancer Res 17:6973–6984
81. Frosina G (2010) The bright and the dark sides of DNA repair in stem cells. J Biomed Biotechnol 2010:845396
82. Rabik CA, Dolan ME (2007) Molecular mechanisms of resistance and toxicity associated with platinating agents. Cancer Treat Rev 33:9–23
83. Wang D, Lippard SJ (2005) Cellular processing of platinum anticancer drugs. Nat Rev Drug Discov 4:307–320
84. Johannessen TC, Bjerkvig R, Tysnes BB (2008) DNA repair and cancer stem-like cells-potential partners in glioma drug resistance? Cancer Treat Rev 34:558–567
85. Di BG, Alessio N, Dell'Aversana C, Casale F, Teti D, Cipollaro M, Altucci L, Galderisi U (2010) Impact of histone deacetylase inhibitors SAHA and MS-275 on DNA repair pathways in human mesenchymal stem cells. J Cell Physiol 225:537–544
86. Bartucci M, Svensson S, Romania P, Dattilo R, Patrizi M, Signore M, Navarra S, Lotti F, Biffoni M, Pilozzi E, Duranti E, Martinelli S, Rinaldo C, Zeuner A, Maugeri-Sacca M, Eramo A, De MR (2012) Therapeutic targeting of Chk1 in NSCLC stem cells during chemotherapy. Cell Death Differ 19:768–778
87. Viale A, De FF, Orleth A, Cambiaghi V, Giuliani V, Bossi D, Ronchini C, Ronzoni S, Muradore I, Monestiroli S, Gobbi A, Alcalay M, Minucci S, Pelicci PG (2009) Cell-cycle restriction limits DNA damage and maintains self-renewal of leukaemia stem cells. Nature 457:51–56
88. Zhuang W, Li B, Long L, Chen L, Huang Q, Liang ZQ (2011) Knockdown of the DNA-dependent protein kinase catalytic subunit radiosensitizes glioma-initiating cells by inducing autophagy. Brain Res 1371:7–15
89. Karimi-Busheri F, Rasouli-Nia A, Mackey JR, Weinfeld M (2010) Senescence evasion by MCF-7 human breast tumor-initiating cells. Breast Cancer Res 12:R31
90. Zhang M, Behbod F, Atkinson RL, Landis MD, Kittrel F, Edwards D, Medina D, Tsimelzon A, Hilsenbeck S, Green JE, Michalowska AM, Rosen JM (2008) Identification of tumor-initiating cells in a p53-null mouse model of breast cancer. Cancer Res 68:4674–4682
91. Fan J, Robert C, Jang YY, Liu H, Sharkis S, Baylin SB, Rassool FV (2011) Human induced pluripotent cells resemble embryonic stem cells demonstrating enhanced levels of DNA repair and efficacy of nonhomologous end-joining. Mutat Res 713:8–17
92. Sarkar FH, Li Y, Wang Z, Kong D (2009) Cellular signaling perturbation by natural products. Cell Signal 21:1541–1547
93. Sarkar FH, Li Y (2009) Harnessing the fruits of nature for the development of multi-targeted cancer therapeutics. Cancer Treat Rev 35:597–607
94. Sarkar FH, Li Y, Wang Z, Padhye S (2010) Lesson learned from nature for the development of novel anti-cancer agents: implication of isoflavone, curcumin, and their synthetic analogs. Curr Pharm Des 16:1801–1812
95. Sarkar FH, Li Y, Wang Z, Kong D (2010) The role of nutraceuticals in the regulation of Wnt and Hedgehog signaling in cancer. Cancer Metastasis Rev 29:383–394
96. Kawasaki BT, Hurt EM, Mistre T, Farrar WL (2008) Targeting cancer stem cells with phytochemicals. Mol Interv 8:174–184

97. Li Y, Wicha MS, Schwartz SJ, Sun D (2011) Implications of cancer stem cell theory for cancer chemoprevention by natural dietary compounds. J Nutr Biochem 22:799–806
98. Ghantous A, Gali-Muhtasib H, Vuorela H, Saliba NA, Darwiche N (2010) What made sesquiterpene lactones reach cancer clinical trials? Drug Discov Today 15:668–678
99. Guzman ML, Rossi RM, Karnischky L, Li X, Peterson DR, Howard DS, Jordan CT (2005) The sesquiterpene lactone parthenolide induces apoptosis of human acute myelogenous leukemia stem and progenitor cells. Blood 105:4163–169
100. Guzman ML, Rossi RM, Neelakantan S, Li X, Corbet CA, Hassane DC, Becker MW, Bennet JM, Sullivan E, Lachowicz JL, Vaughan A, Sweeney CJ, Matthews W, Carrol M, Liesveld JL, Crooks PA, Jordan CT (2007) An orally bioavailable parthenolide analog selectively eradicates acute myelogenous leukemia stem and progenitor cells. Blood 110:4427–4435
101. Kim YR, Eom JI, Kim SJ, Jeung HK, Cheong JW, Kim JS, Min YH (2010) Myeloperoxidase expression as a potential determinant of parthenolide-induced apoptosis in leukemia bulk and leukemia stem cells. J Pharmacol Exp Ther 335:389–400
102. Zhou J, Zhang H, Gu P, Bai J, Margolick JB, Zhang Y (2008) NF-kappaB pathway inhibitors preferentially inhibit breast cancer stem-like cells. Breast Cancer Res Treat 111:419–427
103. Liu Y, Lu WL, Guo J, Du J, Li T, Wu JW, Wang GL, Wang JC, Zhang X, Zhang Q (2008) A potential target associated with both cancer and cancer stem cells: a combination therapy for eradication of breast cancer using vinorelbine stealthy liposomes plus parthenolide stealthy liposomes. J Control Release 129:18–25
104. Kawasaki BT, Hurt EM, Kalathur M, Duhagon MA, Milner JA, Kim YS, Farrar WL (2009) Effects of the sesquiterpene lactone parthenolide on prostate tumor-initiating cells: an integrated molecular profiling approach. Prostate 69:827–837
105. Zuch D, Giang AH, Shapovalov Y, Schwarz E, Rosier R, O'Keefe R, Eliseev RA (2011) Targeting radioresistant osteosarcoma cells with parthenolide. J Cell Biochem (PMID: 22109788; doi: 10.1002/jcb.24002)
106. Gun EJ, Williams JT, Huynh, DT, Iannotti MJ, Han C, Barrios FJ, Kendal S, Glackin CA, Colby DA, Kirshner J (2011) The natural products parthenolide and andrographolide exhibit anti-cancer stem cell activity in multiple myeloma. Leuk Lymphoma 52:1085–1097
107. Volate SR, Kawasaki BT, Hurt EM, Milner JA, Kim YS, White J, Farrar WL (2010) Gossypol induces apoptosis by activating p53 in prostate cancer cells and prostate tumor-initiating cells. Mol Cancer Ther 9:461–470
108. Denissova NG, Nasello CM, Yeung PL, Tischfield JA, Brenneman MA (2012) Resveratrol protects mouse embryonic stem cells from ionizing radiation by accelerating recovery from DNA strand breakage. Carcinogenesis 33:149–155
109. Lu KH, Chen YW, Tsai PH, Tsai ML, Le YY, Chiang CY, Kao CL, Chiou SH, Ku HH, Lin CH, Chen YJ (2009) Evaluation of radiotherapy effect in resveratrol-treated medulloblastoma cancer stem-like cells. Childs Nerv Syst 25:543–550
110. Yang YP, Chang YL, Huang PI, Chiou GY, Tseng LM, Chiou SH, Chen MH, Chen MT, Shih YH, Chang CH, Hsu CC, Ma HI, Wang CT, Tsai LL, Yu CC, Chang CJ (2012) Resveratrol suppresses tumorigenicity and enhances radiosensitivity in primary glioblastoma tumor initiating cells by inhibiting the STAT3 axis. J Cell Physiol 227:976–993
111. Pandey PR, Okuda H, Watabe M, Pai SK, Liu W, Kobayashi A, Xing F, Fukuda K, Hirota S, Sugai T, Wakabayashi G, Koeda K, Kashiwaba M, Suzuki K, Chiba T, Endo M, Fujioka T, Tanji S, Mo YY, Cao D, Wilber AC, Watabe K (2011) Resveratrol suppresses growth of cancer stem-like cells by inhibiting fatty acid synthase. Breast Cancer Res Treat 130:387–398
112. Shankar S, Nal D, Tang SN, Meeker D, Passarini J, Sharma J, Srivastava RK (2011) Resveratrol inhibits pancreatic cancer stem cell characteristics in human and KrasG12D transgenic mice by inhibiting pluripotency maintaining factors and epithelial-mesenchymal transition. PLoS One 6:e16530
113. Li Y, Zhang T, Korkaya H, Liu S, Le HF, Newman B, Yu Y, Clouthier SG, Schwartz SJ, Wicha MS, Sun D (2010) Sulforaphane, a dietary component of broccoli/broccoli sprouts, inhibits breast cancer stem cells. Clin Cancer Res 16:2580–2590

114. Rausch V, Liu L, Kallifatidis G, Bauman B, Mattern J, Gladkich J, Wirth T, Schemmer P, Buchler MW, Zoller M, Salnikov AV, Her I (2010) Synergistic activity of sorafenib and sulforaphane abolishes pancreatic cancer stem cell characteristics. Cancer Res 70:5004–5013
115. Kallifatidis G, Labsch S, Rausch V, Mattern J, Gladkich J, Moldenhauer G, Buchler MW, Salnikov AV, Her I (2011) Sulforaphane increases drug-mediated cytotoxicity toward cancer stem-like cells of pancreas and prostate. Mol Ther 19:188–195
116. Srivastava RK, Tang SN, Zhu W, Meeker D, Shankar S (2011) Sulforaphane synergizes with quercetin to inhibit self-renewal capacity of pancreatic cancer stem cells. Front Biosci (Elite Ed) 3:515–528
117. Zhou W, Kallifatidis G, Bauman B, Rausch V, Mattern J, Gladkich J, Giese N, Moldenhauer G, Wirth T, Buchler MW, Salnikov AV, Her I (2010) Dietary polyphenol quercetin targets pancreatic cancer stem cells. Int J Oncol 37:551–561
118. Tang SN, Singh C, Nal D, Meeker D, Shankar S, Srivastava RK (2010) The dietary bioflavonoid quercetin synergizes with epigallocathechin gallate (EGCG) to inhibit prostate cancer stem cell characteristics, invasion, migration and epithelial-mesenchymal transition. J Mol Signal 5:14
119. Tang SN, Fu J, Nal D, Rodova M, Shankar S, Srivastava RK (2012) Inhibition of sonic hedgehog pathway and pluripotency maintaining factors regulate human pancreatic cancer stem cell characteristics. Int J Cancer 131:30–40
120. Kakarala M, Brenner DE, Korkaya H, Cheng C, Tazi K, Ginestier C, Liu S, Dontu G, Wicha MS (2010) Targeting breast stem cells with the cancer preventive compounds curcumin and piperine. Breast Cancer Res Treat 122:777–785
121. Yu Y, Kanwar SS, Patel BB, Nautiyal J, Sarkar FH, Majumdar AP (2009) Elimination of colon cancer stem-like cells by the combination of curcumin and FOLFOX. Transl Oncol 2:321–328
122. Fong D, Yeh A, Naftalovich R, Choi TH, Chan MM (2010) Curcumin inhibits the side population (SP) phenotype of the rat C6 glioma cell line: towards targeting of cancer stem cells with phytochemicals. Cancer Lett 293:65–72
123. Padhye S, Yang H, Jamadar A, Cui QC, Chavan D, Dominiak K, McKinney J, Banerje S, Dou QP, Sarkar FH (2009) New difluoro Knoevenagel condensates of curcumin, their Schiff bases and copper complexes as proteasome inhibitors and apoptosis inducers in cancer cells. Pharm Res 26:1874–1880
124. Padhye S, Banerje S, Chavan D, Pandye S, Swamy KV, Ali S, Li J, Dou QP, Sarkar FH (2009) Fluorocurcumins as cyclooxygenase-2 inhibitor: molecular docking, pharmacokinetics and tissue distribution in mice. Pharm Res 26:2438–2445
125. Kanwar SS, Yu Y, Nautiyal J, Patel BB, Padhye S, Sarkar FH, Majumdar AP (2011) Difluorinated-curcumin (CDF): a novel curcumin analog is a potent inhibitor of colon cancer stem-like cells. Pharm Res 28:827–838
126. Bao B, Ali S, Kong D, Sarkar SH, Wang Z, Banerje S, Aboukameel A, Padhye S, Philip PA, Sarkar FH (2011) Anti-tumor activity of a novel compound-CDF is mediated by regulating miR-21, miR-200, and PTEN in pancreatic cancer. PLoS One 6:e17850
127. Bao B, Ali S, Banerje S, Wang Z, Logna F, Azmi AS, Kong D, Ahmad A, Li Y, Padhye S, Sarkar FH (2012) Curcumin analog CDF inhibits pancreatic tumor growth by switching on suppressor microRNAs and attenuating EZH2 expression. Cancer Res 72:335–345
128. Lim KJ, Bisht S, Bar EE, Maitra A, Eberhart CG (2011) A polymeric nanoparticle formulation of curcumin inhibits growth, clonogenicity and stem-like fraction in malignant brain tumors. Cancer Biol Ther 11:464–473
129. Lin L, Liu Y, Li H, Li PK, Fuchs J, Shibata H, Iwabuchi Y, Lin J (2011) Targeting colon cancer stem cells using a new curcumin analogue, GO-Y030. Br J Cancer 105:212–20
130. Freitas AA, de Magalhaes JP (2011) A review and appraisal of the DNA damage theory of ageing. Mutat Res 728:12–22
131. Hoeijmakers JH (2009) DNA damage, aging, and cancer. N Engl J Med 361:1475–1485

Chapter 4
DNA Repair in Normal Stem Cells

Olga Momčilović and Gerald Schatten

Abstract Stem cells are self-renewing cells with the ability to differentiate into one or more somatic cell types. Pluripotent stem cells (PSCs) can form any type of somatic cell and play essential roles during development, whereas multipotent stem cells produce a limited number of closely related cell types and are responsible for tissue homeostasis during an organism's lifetime. It is essential for stem cells to maintain genomic integrity, as alterations of their DNA sequence would be transmitted to daughter cells. This might affect developmental processes, tissue cellularity and function, and lead to aging and malignant transformation. The fast pace of the cell cycle in PSCs exposes them to the risk of accumulation of replication errors, whereas the long life of multipotent stem cells predisposes them to accumulation of mutations. PSCs employ several mechanisms to minimize mutational burden, such as low mitochondrial activity and reactive oxygen species (ROS) production, high activity of efflux pumps, expression of telomerase, efficient DNA repair, as well as triggering apoptosis at a lower threshold than other cell types. Quiescence of multipotent stem cells may minimize chances of accumulation of replication errors, and protect them from oxidative stress due to low metabolic activity, mitochondrial respiration, and ROS production. However, quiescence may act as a double-edged sword, since resting in the G_0 phase of the cell cycle limits the choice for DNA double strand break repair to error prone non-homologous end joining.

Abbreviations

DSB	Double strand break repair
ESC	Embryonic stem cell
EpiSC	Epiblast stem cell
ECC	Embryonic carcinoma cell

O. Momčilović (✉)
The North Bay CIRM Shared Laboratory for Stem Cells and Aging,
Buck Institute for Research on Aging, Novato, CA 94945, USA
e-mail: omomcilovic@buckinstitute.org

G. Schatten
Pittsburgh Development Center, Pittsburgh, PA 15222, USA

Department of Obstetrics and Gynecology,
University of Pittsburgh School of Medicine, Pittsburgh, PA 15261, USA
e-mail: schattengp@upmc.edu

L. A. Mathews et al. (eds.), *DNA Repair of Cancer Stem Cells*,
DOI 10.1007/978-94-007-4590-2_4, © Springer Science+Business Media Dordrecht 2013

EGC	Embryonic germ cell
HSC	Hematopoietic stem cell
HR	Homologous recombination
ICM	Inner cell mass
IR	Ionizing radiation
MEF	Mouse embryonic fibroblasts
NHEJ	Non-homologous end joining
NSC	Neural stem cell
PSC	Pluripotent stem cell
ROS	Reactive oxygen species
UV	Ultraviolet radiation

4.1 Introduction

One of the foundations for the existence of life is that genetic information is stable and heritable. Thus, following the discovery of the function of DNA as a carrier of genetic information and the subsequent elucidation of its structure, it became apparent that alterations in DNA could be deleterious for the species. The possible existence of DNA repair mechanisms was initially overlooked and another more intuitive hypothesis was postulated—that the primary structure of DNA is stable and not prone to chemical changes. Only the discovery of DNA secondary structure led to the development of the concept of DNA repair as means of safeguarding genetic information [1].

We now know that DNA is a very metabolically active molecule. It exists in a cellular environment rich in water and reactive oxygen, surrounded by a plethora of metabolic products that can cause hydrolytic and oxidative modifications in DNA bases, phosphodiester bonds, and pentose moieties. Normal products of metabolic processes, such as *S*-adenosylmethionine, can serve as donors of alkylating groups in enzymatic and non-enzymatic reactions, thereby modulating the pairing of DNA bases, whereas errors during DNA replication can change DNA primary structure. The net result of the chemical alterations of DNA building blocks includes base pair transitions, hydrolytic cleavage of N-glycosyl bonds and creation of abasic sites, which collectively create variation in the DNA sequence.

In addition to the endogenous sources of DNA lesions, DNA can be damaged by exogenous (environmental) factors, such as ionizing (IR) and ultraviolet (UV) radiation, as well as a myriad of chemicals and drugs. Exogenous factors can induce DNA damage both directly and indirectly. For example, IR may directly ionize bases and/or deoxyribose groups that absorb radiation energy. However, the most damaging effect of IR is indirect, through radiolysis of water and the production of reactive oxygen species (ROS), such as oxygen superoxide, hydroxyl ions, electrons, and hydrogen peroxide, all of which cause oxidative damage.

Many sources of DNA damage have been present since the beginning of the evolution of life on our planet. Cellular genomes are continuously exposed to DNA damaging agents, therefore in order for life to proliferate, multiple DNA damage response pathways have evolved. Many of these cellular DNA damage defense mechanisms have been studied in somatic cells, and a considerable body of work is now available. However, far less is currently known and understood about the maintenance of genomic integrity in stem cells.

In this chapter, we review DNA repair in normal stem cells. We start our discussion with a description of different stem cell types, followed by a description of DNA damage responses in these cells. We focus on a review of the different DNA repair pathways in pluripotent stem cells (PSCs), followed by brief description of adult stem cells' DNA damage responses.

4.2 Stem Cells

Stem cells are defined by two key properties: (1) the ability to proliferate and create more of the same cell type through the process of self-renewal, and (2) the potential to acquire specialized cellular function(s) through the process of differentiation. All stem cells have the capability to self-renew, but their differentiation spectrums vary considerably, and based on this criterion they are typically categorized as pluripotent or multipotent. PSCs can give rise to the over 200 different cell types present in the human body; they can differentiate into virtually any cell type of an organism. In contrast, multipotent stem cells have restricted differentiation potential and can give rise to a family of closely related cell types.

The zygote and individual cells of the early embryo (more specifically, the two to four cell stage embryos) can form both the embryo proper and extraembryonic tissues and are referred to as being totipotent. At the blastocyst stage, the outer layer of cells, termed trophoblast, forms and contributes to the placenta following implantation of the embryo into the intrauterine wall, whereas the inner cell mass (ICM) gives rise to the embryonic tissues. Although PSCs generate all cell types of the developing organism, they cannot form the extraembryonic tissues necessary for the development of an embryo. Therefore, PSCs are not capable of developing into an organism themselves. During ontogeny, developmental (and differentiation) potential becomes progressively restricted, shifting from totipotent, to pluripotent, to multipotent and, finally, terminally differentiated cells (Fig. 4.1). In this section we will introduce the different kinds of stem cells in greater detail.

4.2.1 Pluripotent Stem Cells

The ability of PSCs to produce any differentiated cell type in an organism makes them invaluable for studying early human developmental stages, which otherwise is unavailable due to ethical and technical concerns. PSCs hold great promise for clinical applications as a source of differentiated cells for cell replacement therapies. However, they can also be used as a tool for generating transgenic animals, and in various other commercial and basic research applications. There are several types

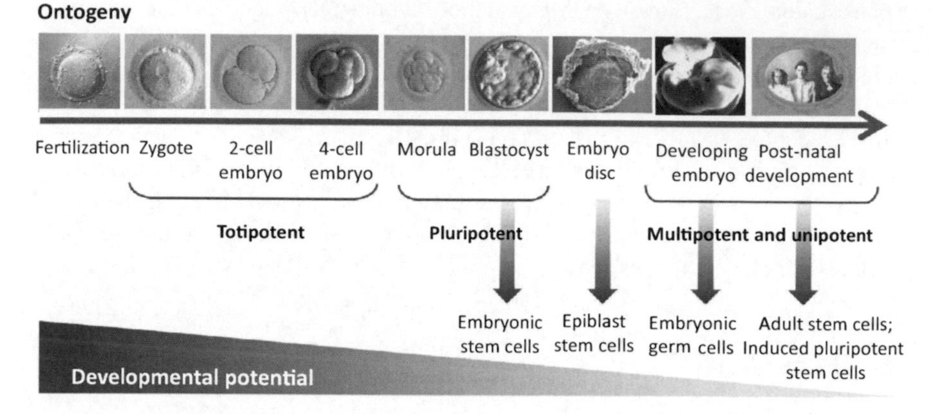

Fig. 4.1 Developmental potential becomes restricted during ontogeny. Following fertilization, a single cell embryo (zygote) is formed. The zygote undergoes cleavage, which increases the cell number unaccompanied by cell growth. The zygote and 2–4 cell stage embryo are totipotent, which means they can form the entire organism. At the blastocyst stage, the outer layer of cells forms the trophoblast, and a small group of cells in the center forms the inner cell mass (ICM), from which embryonic stem cells (ESCs) are derived. During subsequent development, the trophoblast gives rise to the placenta, whereas the ICM produces all tissues of the developing embryo. Note that the ICM is no longer totipotent, since it cannot form the placenta. Epiblast stem cells and embryonic germ cells behave similar to ESCs and are also pluripotent. As development of an organism progresses, the developmental potential of stem cells becomes increasingly restricted, and after birth, only multipotent stem cells exist

of PSCs, including embryonic stem cells (ESC), embryonic carcinoma cells (ECC), embryonic germ cells (EGC), epiblast stem cells (EpiSC), and induced pluripotent stem cells (iPSC) (summarized in Table 4.1). Except for iPSCs, all PSCs are of embryonic origin. Embryonic stem cells are the most studied PSC type, and are used as a gold standard for comparisons with other types of PSCs, however, they were not the first isolated type of PSCs. ECCs were historically the first PSCs available. They were derived from teratocarcinomas (also called teratomas) in the 1960s and 1970s [2, 3]. Teratocarcinomas are usually benign tumors composed of tissues belonging to three germ layers (ectoderm, mesoderm, and endoderm) and an undifferentiated stem cell compartment responsible for the tumor growth. Embryonic carcinoma stem cells are derived from the stem cells present in teratocarcinomas and represent the first known type of cancer stem cells. However, ECCs are genetically unstable and differentiate relatively poorly in comparison to ESCs [4]. Thus, they are not frequently used for studying early development, and are not likely to be used as a source of cells for clinical therapies.

Mouse EGCs are derived from primordial germ cells (PGC), localized in germinal ridges of 9.0–12.5 d.p.c. mouse embryos [5, 6]. Although they share numerous similarities with mouse ESCs, including cellular morphology, expression of pluripotency markers, and comparable contribution to chimeric mice [7, 8], EGCs retain some

Table 4.1 Summary of pluripotent stem cell (PSC) types. The table lists the origins and years of derivation for different PSC types

PSC type	Isolated from	Year	Reference
Mouse embryonic carcinoma cells (ECC)	Teratocarcinomas	1964	[2, 3]
Human embryonic carcinoma stem cells (ECC)	Teratocarcinomas	1977	[13]
Mouse embryonic stem cells (ESC)	Inner cell mass (ICM)	1981	[14, 15]
Mouse embryonic germ cells (EGC)	Primordial germ cells	1992	[5, 6]
Human embryonic stem cells (ESC)	Inner cell mass (ICM)	1998	[16]
Mouse induced pluripotent stem cells (iPSC)	Somatic cells	2006	[17]
Human induced pluripotent stem cells (iPSC)	Somatic cells	2007	[18]
Mouse epiblast stem cells (EpiSC)	Epiblasts	2007	[11, 12]

properties of PGCs, such as global genome demethylation, X-chromosome reactivation, and erasure of imprints that are not present in mouse ESCs [8, 9]. Human EGCs have been isolated from five to seven week old embryos, and can be directed to multilineage development, but have very limited proliferation capacity in *in vitro* culture [10].

EpiSCs are derived from the epiblast (a layer of cells that gives rise to the entire embryo proper) of early mouse and rat post-implantation embryos (E5.5–E6.5) [11, 12]. These cells are almost identical to human ESCs in terms of morphology, pluripotency marker expression, and requirement of growth factors for maintenance of pluripotent state. Mouse EpiSCs can form teratocarcinomas upon injection into immunocompromised mice, but unlike mouse ESCs, poorly contribute to chimeras. Human EpiSCs and EGCs with properties of mouse EGCs have not been isolated to date due to ethical concerns and technical difficulties, but significant effort has been put in place to derive these types of cells from non-human primates.

Embryonic stem cells are extracted from the inner cell mass (ICM) of pre-implantation embryos at the blastocyst stage (about 5 days old human embryo and 2–3.5 d.p.c. mouse embryo) [14–16]. ESCs express proteins present in the ICM, undergo unlimited self-renewal (some human ESCs lines are continually grown in *in vitro* culture for multiple years), and can be directed to differentiate into many cell lineages [14–16]. Mouse studies have proven the differentiation potential of ESCs by injection of mouse ESCs into the blastocyst of the recipient embryo. Injected donor ESCs are able to contribute to the three germ layers and germ line of the resulting chimeric embryo [19]. An even more stringent test of the developmental potential of ESCs is the tetraploid complementation assay. Fusion of cells at the two cell stage mouse embryo produces a tetraploid embryo that cannot develop past mid gastrulation, but can form the blastocyst and extraembryonic tissues. When diploid mouse ESCs are combined with the tetraploid embryo at the morula or blastocyst stage, and transferred into the uterus, normal development proceeds. Tetraploid cells form the extraembryonic tissues, whereas the entire embryo proper is formed by the ESCs, thereby confirming their pluripotency [20–22]. Since these experiments are impossible and unethical in humans, proof of human ESC differentiation potential comes from the injection of ESCs into immunocompromised mice in which they form teratocarcinomas composed of three germ layers.

Table 4.2 Comparison of mouse and human embryonic stem cell (ESC) properties

Property	Mouse ESC	Human ESC
Unlimited self-renewal	Yes	Yes
Pluripotency tested by		
Tetraploid complementation	Yes	No
Chimera formation	Yes	No
Teratoma formation	Yes	Yes
Directed *in vitro* differentiation	Yes	Yes
High nucleus to cytoplasm ratio	Yes	Yes
Tolerance of single cell state (clonal propagation)	Yes	No (require addition of ROCK inhibitor, Y-27632)
Telomerase activity	Yes	Yes
Alkaline phosphatase	Yes	Yes
Core transcriptional regulators of pluripotency		
POU5F1 (OCT4)	Yes	Yes
NANOG	Yes	Yes
SOX2	Yes	Yes
Cell surface markers		
SSEA1	Yes	No
SSEA3	No	Yes
SSEA4	No	Yes
TRA1-60	No	Yes
TRA1-81	No	Yes
Growth factor requirement for self-renewal		
LIF and BMP	Yes	No
FGF2 and TGFβ/ACTIVIN/NODAL	No	Yes
X chromosome activation	Yes	Line-dependent
Imprinting	Commonly lost	Yes

Mouse and human ESCs share numerous similarities, including unlimited self-renewal and broad differentiation potential, but are not identical. For example, mouse and human ESCs differ in cell surface marker expression, growth factor requirements, clonogenicity, genetic, and epigenetic signatures (summarized in Table 4.2). These dissimilarities may reflect species-specific differences between human and mouse early development. However, an alternative hypothesis has been recently postulated: the observed differences may be caused by the slightly different developmental stage at which mouse and human ESCs are derived. As described above, mouse EpiSCs are remarkably similar to human ESCs, and distinct from mouse ESCs [11, 12]. Mouse ESCs and EpiSCs may represent two different pluripotent states: naïve and primed, respectively [23]. With regard to developmental age, human ESCs would then correspond to mouse primed EpiSCs. However, there is an important distinction—mouse EpiSCs are derived from post-implantation embryos, whereas human ESCs are isolated from pre-implantation embryos. Thus, more studies are necessary to establish the true state of pluripotency in human ESCs. Interestingly, "re-wiring" of human ESCs to a naïve pluripotent state by genetic (ectopic expression of OCT4, KLF4, and KLF2) and pharmacological (addition of LIF and inhibitors of GSK3β and ERK1/2) manipulation was achieved recently [24]. The naïve human ESC state seems to be unstable, and cells quickly revert to the primed mouse EpiSC-like state upon withdrawal of expression of genetic factors and chemical inhibitors.

Pluripotency of both mouse and human ESCs is maintained by the core transcription factors, namely POU5F1 (OCT4), NANOG, and SOX2 [25–29]. These master regulators form the transcription network that maintains the undifferentiated state of ESCs and control early development. OCT4, NANOG, and SOX2 bind to promoters of a number of genes and activate expression of genes that promote self-renewal and the undifferentiated state, while repressing expression of genes involved in differentiation. In addition, they promote expression of each other, thereby creating a positive autoregulatory feedback loop [25].

Multiple studies have shown that cellular identity may be changed by forced expression or repression of key transcription factors that specify cell fate [30–32]. This suggests that with the understanding of the master regulatory transcription network in ESCs, it may be possible to "reprogram" fully differentiated somatic cells into an ESC-like state. Furthermore, reprogramming of somatic cells into ESC-like cells has been achieved previously by somatic cell nuclear transfer (SCNT) [33], as well as by fusion of somatic cells with PSCs [9, 34]. These experiments made it clear that oocytes and PSCs contain factors that can reprogram somatic cellular genomes into a pluripotent state. Yamanaka and his team successfully reprogrammed mouse somatic cells into a pluripotent-like state by overexpressing a set of four genes, namely, OCT4, KLF4, SOX2, and cMYC [17]. The resulting cells form ESC-like colonies, express ESC markers, contribute to all three germ layers in a mouse chimera, and are therefore termed induced pluripotent stem cells (iPSC). This finding was transferred to human somatic cells a year later, when two independent teams reprogrammed human somatic cells into iPSCs using two different cocktails of transcription factors. Yamanaka's team used the same combination of transcription factors as for reprogramming of mouse cells to enable human cell reprogramming [18], whereas Thomson and colleagues used OCT4, NANOG, SOX2, and LIN28 [35] to achieve the same goal. Since these pioneering experiments, iPSCs have been derived from numerous species, using refined protocols, as well as from patients with various genetic diseases [36–42].

4.2.2 Multipotent Stem Cells

Multipotent stem cells share with PSCs the ability to self-renew. However, multipotent stem cells have restricted differentiation potential in comparison to PSCs, and can give rise only to several closely related cell types. For example, neural stem cells (NSCs) can give rise to neurons and glial cells of the nervous system, whereas hematopoietic stem cells (HSCs) can differentiate into any blood cell type. However, NSCs cannot contribute to blood, and HSCs do not give rise to nervous tissue.

All tissues of an organism are exposed to varying degrees of "wear and tear" and depend on the action of multipotent stem cells for the replacement of dead and

Table 4.3 Comparison of pluripotent and multipotent stem cells' properties

	Pluripotent stem cells	Multipotent stem cells
Differentiation potential	All cell types	Restricted; related cell type
Self-renewal capacity	Unlimited in cell culture	Limited in cell culture
Life span *in vivo*	Short lived; several days	Long lived; multiple years
Origin	Embryonic	Embryonic and adult
Function	Generate all cell types	Maintenance of tissue homeostasis

damaged cells in order to maintain tissue cellularity and function. Since multipotent stem cells are specialized to a certain extent, every organ system has its own set of multipotent stem cells (a few examples being HSCs, NSCs, muscle stem cells, corneal stem cells, skin stem cells, bone stem cells, mesenchymal stem cells, and spermatogonial stem cells). Different tissues have varying level of regenerative potential, which is reflected by the proliferative capacity of resident stem cells. For example, in high turnover tissues, such as the hematopoietic system, proliferation and differentiation of HSCs into specialized progeny has been well documented. In contrast, in organs with lower regenerative potential, such as heart or brain, identification of resident stem cells has been more challenging [43].

Unlike ESCs, whose *in vivo* counterpart, ICM, exists for very short period of time during embryonic development, multipotent stem cells are present in organs after birth, and are frequently referred to as adult stem cells (a comparison between ESCs and adult stem cells is summarized in Table 4.3). They reside in a specialized niche that is composed of supporting cells and extracellular molecules, and their long-term maintenance depends to a large extent on interaction between stem cells and their niche.

Under optimal physiological conditions, most adult stem cells reside in the G_0 stage of the cell cycle. In the presence of mitogenic signals stem cells re-enter the cell cycle and undergo asymmetrical division, giving rise to one stem cell and one progenitor, or transient-amplifying (TA) cell (Fig. 4.2) [44]. TAs undergo several rounds of division, thereby expanding the pool of differentiating cells. Adult stem cells can switch to symmetrical cell division giving rise to two stem cells at times when the stem cell pool needs to be expanded, such as during growth or wound healing. In this hierarchical model, adult stem cells are rare, quiescent cells responsible for tissue repopulation, and are greatly outnumbered by the dividing progenitor population from which differentiated cells arise [45].

Adult stem cells are vital for organ function throughout the life of an organism. Therefore, it appears intuitive that the existence of a pool of self-renewing multipotent stem cells with robust potential for differentiation into specialized progeny is of utmost importance for an organism's optimal health and survival. Another implication is that impaired function and depletion of adult stem cells with aging or disease will lead to reduced proliferative response, misdirected differentiation, and overall impaired regeneration of an organ. Finally, these long-lived, self-renewing stem cells may accumulate mutations during their life time and represent a preferred cellular compartment for malignant transformation [43, 46].

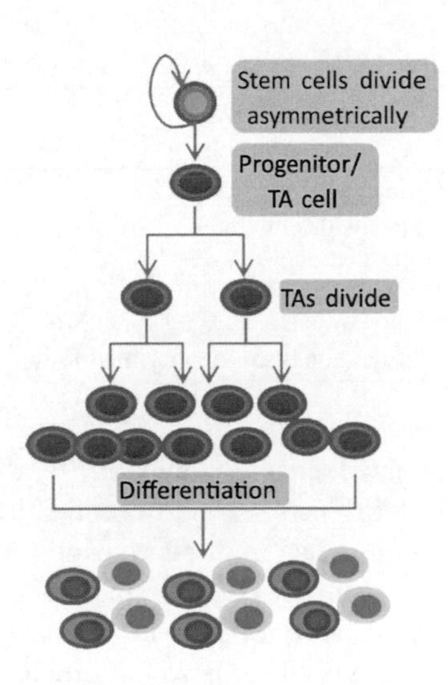

Fig. 4.2 Asymmetric division of multipotent stem cells. Most of the multipotent stem cells are rare, quiescent, tissue specific stem cells. When drawn into the cell cycle, multipotent stem cells divide asymmetrically producing one stem cell and one progenitor cell. Progenitors undergo several cellular divisions thus amplifying the pool of differentiating cells

4.3 DNA Damage in Stem Cells

The continuous challenge of cellular genomes with genotoxic stressors has acted as an evolutionary pressure for organisms to develop protective mechanisms to counteract these challenges. The two main strategies for the limitation of DNA alterations and maintenance of pristine cellular populations are: (1) the suppression of mutation, and (2) the elimination of cells that have acquired DNA mutations. The very metabolic processes that keep us alive also produce a plethora of genotoxic stressors that introduce a significant DNA damage burden in each cell every day (Table 4.4). However, most of these lesions do not become permanent alterations of DNA sequence due to mutation suppression mechanisms. DNA lesions normally activate cell cycle checkpoints that arrest cell cycle progression and provide time for DNA repair mechanisms to remove DNA damage, preventing the transmission of genetic changes to daughter cells. However, if DNA damage cannot be repaired, or overwhelms cellular repair capacity, cells may exit the cell cycle permanently and senesce or undergo cell death. In cases when cells fail to repair DNA damage, or exit the cell cycle, genetic alterations become permanent and are transmitted to daughter cells in the next round of replication. Most mutations do not have deleterious effects, but given the size of the genome, and number of replications through the ontogeny, mutations do accumulate and eventually impair cellular function leading toward disease. Thus, it is essential for cells to safeguard their genomic integrity in order to prevent mutagenesis, as well as to avoid extensive cell death.

As discussed in the introduction, “DNA damage is an inescapable aspect of life” [1], and stem cells are not spared from DNA lesions. Since stem cells produce differentiated cells, they face unique challenges in maintaining the fidelity of genetic information. For example, cells that comprise ICM, and from which ESCs are

Table 4.4 Endogenous DNA damage frequencies [1]

Type of DNA damage	Frequency (number of lesions per cell per 24 hours)
Oxidation of guanosine (8-oxoG)	1,000–2,000
Cytosine deamination	100–500
Depurination	18,000
Depyrimidination	600
Methylation by *S*-adenosylmethionine (7-Methylguanine)	6,000
Methylation by *S*-adenosylmethionine (3-Methyladenine)	1,200

isolated, undergo rapid proliferation. This may expose them to accumulation of mutations due to errors in replication, and/or replicative stress. Mutations in the ICM would affect multiple lineages, and lead to major disruption in the development of an organism, resulting in birth defects and death. Therefore, these cells may favor maintenance of genomic integrity over cell survival. In contrast, mutations in multipotent stem cells would affect a limited number of cell lineages. Since these cells are responsible for tissue homeostasis throughout an organism's lifetime, they need to balance between mechanisms that safeguard genomic integrity and their ability to self-renew and differentiate. Elimination of stem cells with mutations would maintain pristine cell population, but would eventually lead to the depletion of the stem cell pool.

A great body of literature describes DNA damage responses (DDR) in somatic cells, but far less is known about DDR in stem cells. In the following sections we will focus on repair of DNA damage by stem cells. We will first discuss DNA repair in pluripotent stem cells (ESCs and iPSCs), followed by an overview of DDR in multipotent adult stem cells.

4.3.1 DNA Repair in Pluripotent Stem Cells

DNA repair is a part of the cellular DDR, which also includes checkpoint activation, cell death, and senescence. Therefore, in order to put in context findings about DNA repair in the ESCs and iPSCs, we first need to describe some of their unique DDR.

ESCs are primary immortal cells that do not undergo senescence or contact inhibition in culture. Mouse, nonhuman primate, and human ESCs proliferate rapidly, similar to cells in the early embryos; ESCs have an abbreviated cell cycle, mainly due to shortening of the G_1 phase [47–51]. In mouse ESCs, most of the cell cycle regulators are expressed in a cell cycle independent manner [47–49, 52]. In contrast, in human ESCs, most of the regulators demonstrate phase-specific activity (for detailed review of cell cycle regulation in ESCs see [53]).

The vast majority of ESCs (50–60 %) are in the S phase of the cell cycle [50, 54]. In healthy ESC cultures, all cells are positive for the proliferative marker Ki-67, demonstrating the absence of quiescent cells [50]. ESCs of different species exhibit the absence of functional G_1/S cell cycle checkpoint and arrest in the G_2/M phase of the cell cycle following DNA damage [54–57]. The absence of G_1/S arrest in ESCs suggests that cells with damaged DNA may enter the S phase when DNA damage would be exacerbated and promote apoptosis. Indeed, both mouse and human ESCs are hypersensitive to genotoxic insults and undergo apoptosis within hours of exposure to DNA damage [54, 55]. The lack of G_1/S cell cycle arrest and DNA damage hypersensitivity in mouse ESCs have been attributed to aberrant localization of CHEK2 and TP53 to the cytoplasm following DNA damage where they are unable to perform their normal function [55]. However, in human ESCs and iPSCs these checkpoint signaling molecules are phosphorylated and localized to the nucleus where they appear to perform their normal function [54]. One group has recently demonstrated that following UV irradiation of synchronized human ESCs in G_1 phase, a G_1/S block is established, TP53 is phosphorylated and localized to the nucleus, but CDKN1A (p21) protein level remains the same (although mRNA level increases) [58]. At the molecular level, G_1/S arrest appears to be executed by CHEK1/CHEK2 dependent CDC25A degradation, rather than CDKN1A activation. However, G_1/S arrest was not detected following ionizing radiation of asynchronous human ESCs, although TP53 and CHEK2 phosphorylation was evident [54, 57]. This discrepancy remains to be elucidated. Finally, DNA damage induced differentiation of ESCs has been proposed as a unique response of ESCs to radiomimetic drug-induced DNA lesions. Under this hypothesis, cells with damaged DNA differentiate and are therefore eliminated from the pool of cycling PSCs [59, 60]. This hypothesis is still being tested, particularly in light of newer studies that show continuous expression of pluripotency markers and no loss of differentiation potential (our unpublished results and [61]) and pluripotent marker expression after acute DNA damage [54, 61, 62]. It is possible that radiomimetic drugs induce differentiation by continuous infliction of DNA damage and persistent DNA damage signaling. Comparison of DDR between somatic cells and ESCs is summarized in Fig. 4.3.

Multiple mechanisms that maintain genomic integrity are active in ESCs, and include G_2/M cell cycle arrest, efficient DNA repair, and enhanced stress defense pathways. Overall, both mouse [63] and human [64] ESCs have increased capacity for DNA repair in response to various genotoxic treatments, which will be discussed below. We and others have shown that human ESCs and iPSCs have elevated expression of genes that encode proteins that participate in multiple DNA repair pathways [62, 64]. During differentiation of mouse [63] and human [65] ESCs, expression of antioxidant and DNA repair genes are reduced, while the frequency of γH2AX (a marker of DSBs) positive cells increases. High levels of antioxidant proteins in ESCs can neutralize the toxic effect of ROS, whereas P-glycoproteins acting as efflux pumps can eliminate toxic chemicals. Finally, ESCs express both RNA and protein components of telomerase (TR and TERT) that maintain the length of telomeres during ESC proliferation. Following differentiation, mouse ESCs acquire G_1/S checkpoint arrest, mitochondrial mass and ROS production increase, whereas expression of antioxidant enzymes and TR and TERT decreases [55, 62, 63, 65]. Therefore, the strategies for maintenance of genomic integrity in ESCs are dramatically different from those employed by somatic cells.

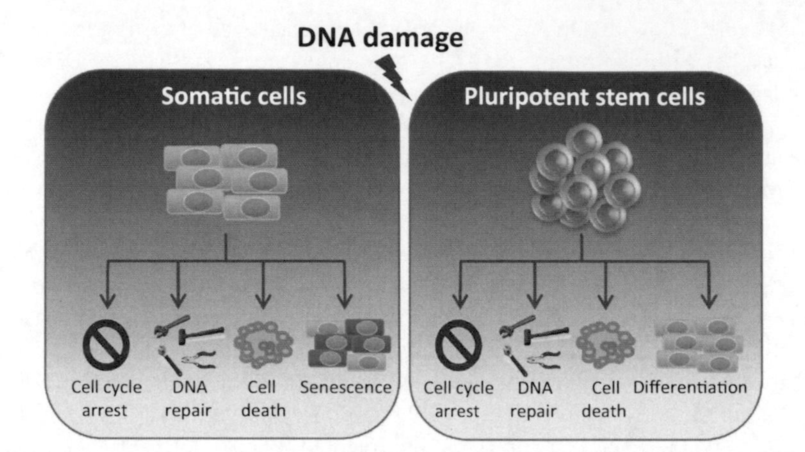

Fig. 4.3 DNA damage responses in somatic cells and pluripotent stem cells (PSCs). DNA damage evokes numerous cellular responses in somatic cells: checkpoint signaling arrests cell cycle progression allowing time for repair of DNA. When DNA damage is excessive, cells may undergo apoptosis, or exit the cell cycle permanently and become senescent. DNA damage responses in PSCs are far less understood, but the data demonstrate activation of DNA damage signaling and cell cycle arrest. DNA repair of at least some types of DNA lesions is more efficient than in somatic cells, although the threshold for apoptosis is very low. Differentiation is another possible response to the presence of DNA damage in PSCs

4.3.1.1 Double Strand Break Repair

Double strand breaks (DSBs) are the most difficult type of DNA damage to repair due to the loss of integrity of both DNA strands. Therefore, DSBs are extremely toxic for cells and even one DSB can be potentially lethal [66]. It is not surprising, then, that cells have evolved several DSB repair pathways, with two main pathways being: homologous recombination (HR) and non-homologous end joining (NHEJ).

HR is critically dependent on the presence of the homologous regions because it uses the genetic information on undamaged sister chromatid as a template to restore the original sequence of a broken DNA molecule. Consequently, it is predominant in the late S and in G_2 phases of the cell cycle when the duplicated chromatids are present [67, 68]. The homologous chromosome can be used as a template during the G_1 phase, but this would result in loss of heterozygosity, which can be more deleterious than error-prone NHEJ.

NHEJ is an error-prone repair process that involves processing of DNA ends which may lead to loss of nucleotides on both ends of a DSB [69–71]. It does not require the presence of the sister chromatid in the cell as a template for repair; hence, it is active in all phases of the cell cycle, and represents the main repair pathway during G_1 and early S phases [72, 73]. NHEJ also has a physiological role during normal development of the immune system where it is essential for V(D)J recombination in immunoglobulin genes. Mutations in genes that encode NHEJ proteins lead to severe combined immunodeficiency (SCID) and increased radiosensitivity.

The choice of DSB repair pathway depends on the cell cycle stage, source of DSBs, kinetics of repair, and cell type. The majority of cells in an organism are post-mitotic, hence a sister chromatid is not present, and cells rely mainly on NHEJ for DSB repair. Furthermore, the availability of pathway-specific proteins regulates the choice of pathway: when NHEJ-specific proteins are absent, HR is stimulated, and when NHEJ proteins are overexpressed, HR is repressed [74]. *In vivo* analysis of recruitment of DNA repair factors to the sites of laser-induced DSBs reveal temporary localization of NHEJ components in an attempt to immediately repair DSBs, followed by prolonged occupancy by HR proteins [75]. Association of RAD51, a primary eukaryotic recombinase, with chromatin during the S phase and its role in repair of replication fork collapse induced DSBs has been well documented [76]. Therefore, DSBs associated with errors during replication are primarily repaired by HR, whereas NHEJ plays a pivotal role in radiation induced DSB repair [68, 77–79].

DSB repair is thus far the most studied form of DNA damage in ESCs because of the lethality of a single DSB. ESCs are particularly exposed to the risk of formation of DSBs at the sites of collapsed replication forks due to rapid proliferation; in fact, the predominant cause of DSBs in proliferating cells are errors during replication, which are typically repaired by HR as discussed above. When assessing the relative roles of HR and NHEJ in DSB repair in ESCs, it appears intuitive that HR would be the preferential mechanism: at any point of time, the majority of ESCs (more than 80 % of cells) are in the S or G_2/M phases of the cell cycle when the sister chromatid is present in the cell [50, 54]. It also appears intuitive that there would be an evolutionary selection mechanism against PSCs that have accumulated incorrectly repaired DNA. Mutation in the DNA of ESCs would have detrimental consequences, not only for an organism, but for the entire species, since they would impair preservation of the species identity. Thus, the principal hypothesis is that HR plays pivotal role in DSB repair in ESCs.

The initial insight into the roles of HR and NHEJ during development came from mouse knock out studies. In general, mice with a complete lack of RAD51, MRE11, NBS1, and RAD50 are early embryonic lethal [80]. RAD51, RAD50, MRE11, and NBS1 null ES cells cannot be isolated from knock out mouse embryos, nor generated *in vitro*, suggesting that the MRN complex and RAD51 are required for normal cellular function, proliferation, and growth [81–84]. RAD52 and RAD54 knock-out mice are viable and show no impaired viability, fertility, or immune system deficiency, suggesting that RAD52 and RAD54 are not essential for normal mouse development [85]. Mouse RAD52 nullizygous ES cells do not show signs of radiosensitivity, and exhibit a 30–40 % reduction in homologous recombination, unlike yeast mutant Scrad52 cells which are not viable [86]. RAD54 inactivation in mouse ES cells leads to decreased homologous recombination and increased radiosensitivity. Together, these results indicate that there are functionally related genes in mammalian cells that can compensate for the absence of RAD52 and RAD54.

Mice lacking any of the genes involved in NHEJ are either viable, or die late in embryonic development. Nullizygous LIG4 (Ligase IV) and XRCC4 mice are late embryonic lethal due to massive TP53 dependent neuronal apoptosis, arrested lymphogenesis and other cellular defects [87–89]. TP53 null background rescues embryonic lethality, but not lymphocyte development or radiosensitivity. XRCC5 (KU80) or XRCC6 (KU70) inactivation in mice results in growth retardation, profound immunodeficiency, as well as marked radiosensitivity and inability to perform end-joining at the cellular level [90–93]. Mice lacking PRKDC (DNA-PKcs) are immunodeficient, but do not exhibit growth retardation [90].

Several research groups demonstrated differences in expression of HR and NHEJ proteins and repair capacities between mouse ESCs and somatic cells (Fig. 4.4). For example, expression of multiple proteins involved in HR, such as members of the RAD52 epistasis group (RAD51, RAD52, and RAD54), were elevated in mouse ESCs relative to MEFs [94, 95]. At the same time, the expression of NHEJ proteins appeared more varied: KU70 and KU80 were elevated, whereas LIG4 and DNA-PKcs were downregulated in mouse ESCs in comparison to fibroblasts [94–96]. Use of HR and NHEJ reporter plasmids enabled Tichy et al. to show that mouse ESCs possess a greater capacity to repair DSBs by HR than somatic cells. When ESCs were induced to differentiate, NHEJ became the predominant DSB repair pathway, whereas HR was significantly reduced [95]. Similar results were obtained by Serrano et al. using *in vitro* HR and NHEJ assays with whole cell extracts [94]. These authors also investigated localization of RAD51 during the cell cycle. Localization of RAD51 to the foci is commonly used as a surrogate marker for HR, and coupled with incorporation of nucleotide analogs BrdU or EdU, allows for the following of RAD51 distribution during the cell cycle. These experiments revealed that in fibroblasts, RAD51 localized to the foci only during the S phase (presumably assisting in collapsed replication fork recovery) so that only 4 % of RAD51 foci were found in fibroblasts outside the S phase, whereas in mouse ESCs, RAD51 foci were found in 58 % of cells outside the S phase. Following cell cycle synchronization, RAD51 foci were present in 68 % of ESCs in the G_1 phase in the absence of exogenous DNA damage, and in 100 % of G_1 cells after ionizing irradiation. Finally, FACS analysis confirmed non-phasic RAD51 expression during all stages of the cell cycle in mouse ESCs, unlike in fibroblasts where it peaked during the S phase. Therefore, RAD51 is recruited to DSBs that originate from replication fork collapse, as well as from other sources in mouse ESCs [94].

Collectively, these studies overwhelmingly document role of HR in repair of DSBs during all stages of the cell cycle in mouse ESCs. Nevertheless, NHEJ activity still contributes to mouse ESCs' survival, as demonstrated by studies in wild type, $H2AX^{-/}$, $ATM^{-/-}$ and $DNA\text{-}PKcs^{-/-}$ mouse ESCs [96]. Mouse ESCs defective in ATM or H2AX exhibited faster DSB repair, higher level of DNA-PKcs, and inability to form foci of phosphorylated ATM (although ATM was autophosphorylated) in comparison to wild type mouse ESCs. In addition, mutant H2AX and DNA-PKcs mouse ESCs were more radiosensitive than wild type cells. Inhibition of DNA-PK activity resulted in reduced DSB rejoining in H2AX knock-out, but not in wild type mouse ESCs, suggesting that DNA-PK has an important role in DSB repair in H2AX deficient mouse ESCs. In wild type mouse ESCs, DNA-PKcs activity still contributed to their survival following irradiation, because inhibition of DNA-PKcs with its specific inhibitor, NU7026, reduced survival of both wild type and H2AX

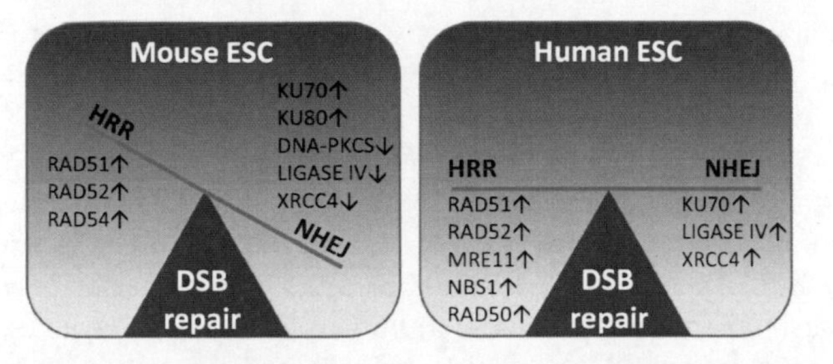

Fig. 4.4 Relative contributions of homologous recombination (HR) and non-homologous end joining (NHEJ) to double strand break repair (DSB) and expression of pathway-specific proteins in embryonic stem cells (ESCs). Mouse ESCs exhibit increased expression of HR-specific proteins and depend mostly on HR for DSB repair. In human ESCs, both pathways appear to be active and expression of both HR- and NHEJ-specific proteins is elevated in comparison to differentiated cells

mutant mouse ESCs. Taken together, it appears that DNA-PK dependent DNA repair is not the main pathway for repair of DSBs, but still contributes to survival of mouse ESCs following DNA damage. In instances when homologous recombination is impaired, mouse ESCs can redirect DSB repair toward NHEJ [96].

Finally, although mouse ESCs possess a low capacity to repair DSBs by NHEJ, and utilize HR for repair of DSBs in all stages of the cell cycle, they still express high level of KU70 and KU80 [95, 96]. The level of these proteins was found to reduce during differentiation. It is possible that KU proteins perform functions other than NHEJ in mouse ESCs, such as maintenance of telomere ends (prevention of end-to end fusion of chromosomes), assist in loading of pre-replication complexes to the origins of replication, or inhibition of apoptosis [95, 96]. However, these functions still need to be confirmed.

Several groups have demonstrated that human ESCs have fewer γH2AX foci per cell relative to differentiated cells [65, 97], and that they have increased capacity to repair multiple forms of DNA damage [64]. We and others have documented that similar to the findings in mouse ESCs, human ESCs and iPSCs show elevated expression of HR repair proteins (summarized in Fig. 4.4), including RAD51, RAD52, RAD50, MRE11, and NBS1, relative to somatic cells [62, 97]. We have also shown that both human ESCs and iPSCs efficiently repair DSBs, as revealed by removal of ionizing radiation induced γH2AX foci within 4–6 hours of DNA lesions. Furthermore, these DSBs are removed at least in part by HR, as evidenced by formation of RAD51 foci and sister chromatid exchanges after DNA damage [62]. Another group has confirmed these findings. Adams et al. demonstrated that a high percentage of isogenic human ESCs and neural progenitors (NPs) form RAD51 foci after irradiation (75 and 65 %, respectively, 6 hours post-irradiation), whereas only 3 % of astrocytes contained RAD51 foci. HR appeared to be ATR dependent, since ATR siRNA reduced formation of γH2AX foci by 70 % in irradiated human ESCs. In contrast, DSB repair in astrocytes was largely dependent on ATM, and ATM inhibition strongly reduced the number of γH2AX foci in these cells, unlike in human ESCs [98].

Expression of the NHEJ proteins XRCC4 and LIG4 were found to be higher in human ESCs and iPSCs, whereas DNA-PKcs and KU70 protein levels were similar to those in somatic cells [62, 96, 97]. In comparison to mouse ESCs, however, expression of multiple NHEJ proteins was significantly elevated [62, 97]. Interestingly, following ionizing irradiation, a proportion of γH2AX foci co-localized with RAD51, and a proportion of γH2AX foci co-localized with KU70, indicating that both HR and NHEJ were active in human ESCs and iPSCs [97]. An *in vitro* assay for DSB repair using linearized plasmid and human ESC or iPSC whole cell extracts, followed by sequencing of DNA junction in repaired plasmids, unveiled that both cell types exhibited an elevated efficiency of DSB repair relative to differentiated cells [97]. The repaired junctions in recovered plasmids were relatively accurate, with loss of a few nucleotides at the broken DNA ends in ESCs and iPSCs, in contrast to differentiated counterparts in which deletion of more than 20 nucleotides was frequently detected [97]. One group recently reported that DSB repair efficiency depends on both chromosomal location of the single DSB and the stage of differentiation [99]. This group utilized zinc-finger nucleases (ZFN) to target double-repeat (DR)-GFP reporter plasmids to different genomic loci. The plasmids carried two non-functional copies of GFP, each interrupted by a recognition site for the I-SceI restriction enzyme. Following transduction of the restriction enzyme into cells, only accurate resolution by HR would restore the GFP gene, which could be measured by FACS. Relative contribution of NHEJ and single strand annealing could be estimated by sequence analysis of the genomic locus carrying the reporter plasmid. These studies demonstrated that DSB repair efficiency can vary 20–260-fold between different chromosomal sites in isogenic human ESC lines. Importantly, the accuracy of DSB repair was elevated in human ESCs relative to differentiated cells, suggesting that the contribution of DSB repair pathways dramatically change during development [99].

While the role of accurate NHEJ in repair of DSBs in human ESCs has been well documented, the role of DNA-PKcs is still unclear. In two recent papers Adam et al. demonstrated that DNA-PKcs were dispensable for accurate NHEJ [98, 100]. These authors have engineered a human ESC line with lentivirus carrying a repair cassette in which DsRed start ATG codon was preceded by an out of frame ATG flanked by two I-SceI recognition sites. Following transduction of adenovirus expressing I-SceI, decoy ATG was excised and after NHEJ repair, DsRed was expressed from its start codon. The repair of DSB was measured by FACS, and accuracy was revealed by sequencing, which determined the extent of modifications at the repair site. During differentiation of human ESCs to NPs, little change in NHEJ efficiency was observed, whereas NHEJ kinetics was significantly increased (2.6-fold) in terminally differentiated astrocytes. The accuracy of repair decreased during differentiation of human ESCs to NPs (1.4-fold), and to astrocytes (2.6-fold). Furthermore, NHEJ was unaffected by inhibition of ATM and DNA-PKcs in ESCs; in contrast, NPs and astrocytes exhibited progressively increasing dependency on ATM and DNA-PKcs for DSB repair. Elimination of XRCC4 greatly impaired efficiency of NHEJ, suggesting that repair of I-SceI induced DSB depended on XRCC4 and canonical NHEJ in human ESCs. Thus, it appears that human ESCs are capable of executing accurate NHEJ of DSBs, and that efficiency of repair by NHEJ increases during differentiation with concomitant loss of accuracy [100].

Interestingly, there is a general difference between mouse and human somatic cells in their dependence on NHEJ. For example, DNA-PK activity [101] and KU70 expression [102] are higher in human relative to mouse somatic cells, and KU80 is an essential protein in human, but not mouse cells [103]. In addition, human ESCs rejoin DSBs faster than mouse ESCs [96] suggesting there might be a difference between mouse and human ESCs' choice of repair pathway.

More recently, a novel DSB repair pathway termed microhomology mediated end joining (MMEJ), or backup NHEJ (B-NHEJ) has been described. This DSB repair mechanism leads to deletion of multiple nucleotides at the both ends of the break, until repeat sequences are encountered and utilized in repair. Therefore, MMEJ is highly mutagenic, unlike NHEJ, which depends on activities of DNA-PKcs, LIG4 and XRCC4, MMEJ relies on poly (ADP-ribose) polymerase 1 (PARP-1), Ligase III/XRCC1 and histone H1 for DSB repair. When the level of these proteins was quantified, Ligase III, XRCC1, and PARP-1 were found to be elevated in mouse ESCs relative to MEF. However, the measured activity of MMEJ was found to be similarly low in both cell types, suggesting that mouse ESCs and differentiated cells did not utilize MMEJ frequently for repair of DSBs [95]. In human ESCs, MMEJ also did not appear to play a major role in DSB repair, since inhibition of PARP-1 had no effect on repair of I-SceI induced DSBs [100].

In aggregate, multiple studies strongly demonstrate that human ESCs and iPSCs utilize both HR and accurate form of NHEJ pathways for repair of DSBs, whereas mouse ESCs rely mostly on HR (Fig. 4.4).

4.3.1.2 Nucleotide Excision Repair

Nucleotide excision repair (NER) is involved in the removal of bulky lesions that cause local distortion in DNA molecules. The most studied NER substrates are UV-induced photoproducts cyclobutane pyrimidine dimers (CPD) and pyrimidine-pyrimidone (6–4) adducts (6–4 photoproduct). NER is initiated by recognition of a distorted DNA helix, followed by unwinding of strands surrounding DNA damage, and excision of 24–32 long oligonucleotides containing the altered nucleotide. Excision is followed by repair synthesis of DNA to fill in the gap, and the repair is completed by ligation of the remaining nick. In cases where DNA damage occurs in the transcribed strand of the active gene, a variation of NER, termed transcription-coupled nucleotide excision repair (TC-NER), is employed. TC-NER enables more efficient (and faster) repair of a transcribed DNA strand. It is worth mentioning that TC-NER repairs the same kind of bulky lesions as "global" NER (global genome NER, GG-NER), but requires additional factors to couple NER to the transcriptional machinery.

Early studies of NER in mouse ESCs showed that within a 24 hour period, CPD could not be removed from transcribed or non-transcribed strands of active genes, or from inactive genes. Clearing of 6–4 photoproducts reached maximum at 30 % for both strands within a 4 hour repair period. Measurement of unscheduled DNA synthesis (UDS), which is commonly used as an indicator of DNA repair, revealed that saturation of the repair capacity occurred at a much lower dose in ESCs (5 J/m^2) relative to MEFs (15 J/m^2); only at a very low dosage of UV irradiation (2.5 J/m^2) was the level of repair synthesis in ESCs similar to the one in MEFs. The role of NER in protection from UV-induced cytotoxicity was confirmed by comparing survival responses of wild type and NER defective ($ERCC1^{-/-}$) ESCs, when the latter exhibited two fold lower survival to the same dose. Thus, NER has a protective effect in ESCs against DNA damage induced with lower dosage of UV, whereas at doses above 5 J/m^2 the repair capacity becomes saturated and cells undergo apoptosis [104].

De Waard et al. compared sensitivity of mouse ESCs and MEFs carrying mutations in various components of the NER pathway: $CSB^{-/-}$ (TC-NER), $XPC^{-/-}$ (GG-NER), and $XPA^{-/-}$ (total NER) [105]. The most sensitive to UV killing were cells with a total defect in NER ($XPA^{-/-}$ ESCs and MEFs). However, the relative contribution of GG-NER and TC-NER was different between ESCs and MEFs: greater sensitivity was observed in XPC-deficient ESCs than XPC-mutant MEFs, whereas in fibroblasts, CSB deficiency sensitized cells to UV-induced apoptosis to a greater extent than in ESCs. Thus, UV sensitivity in MEFs were predominately due to a TC-NER defect, whereas in ESCs it appeared to be caused by GG-NER deficiency. $XPC^{-/-}$ and surviving $XPA^{-/-}$ ESCs showed delayed progression through the S phase and accumulation in G_2/M, which may prevent heavily damaged ESCs from cycling. Further experimentation revealed that CSB and XPA mutant ESCs were sensitive to Illudin S, a drug that causes damage reparable by TC-NER, and that TC-NER, but not GG-NER, deficiency caused an elevated mutation rate following UV irradiation of ESCs, indicating that TC-NER is functional in ESCs.

Human ESCs were reported to have more efficient NER than fibroblasts [64]. The expression of several genes that encode NER proteins (ERCC2, RPA1, and XRCC1) were found to be elevated in human ESCs and iPSCs relative to differentiated cells. However, the expression of ERCC1, XPA and XPC was very similar [62]. No studies on the efficiency of NER in iPSCs have been reported thus far.

Taken together, these results suggest that TC-NER is operational in ESCs but that its effects may be "masked" by overwhelming DDR in entire genome. Finally, it is possible that GG-NER would be critical for the maintenance of the entire genome, rather than just transcribed genes in ESCs, since ESCs differentiate into all cell lineages of an organism, each with a distinct transcriptional program. Reliance on TC-NER in ESCs may be detrimental because it would allow accumulation of mutations in non-transcribed genes that may become activated later in development [105].

4.3.1.3 Base Excision Repair

Base excision repair (BER) is involved in removal of modified DNA bases, such as 7-methyl-purines, 3-methyl-purines, 8-oxoguanine, O^6-alkylguanine, and of uracil. During BER, DNA glycosylases cleave the N-glycosyl bond, releasing the base and creating an abasic site (AP). AP endonucleases hydrolyze phosphodiester bonds 5′ to the AP site, creating a deoxyribose-phosphate that is removed by exonuclease. Finally, the removed nucleotide is replaced and the nick is covalently sealed.

One study demonstrated that mouse ESCs express high levels of multiple BER proteins (APE1, DNA ligase III, UNG2, XRCC1). Mouse ESCs appeared more proficient than MEF in repair of DNA templates containing uracil in *in vitro* DNA incision and DNA incorporation assays using nuclear extracts [106]. These findings parallel those in human ESCs. Following exposure to H_2O_2 and analysis by comet assay 6 hours after oxidative DNA damage, human ESCs had significantly shorter comet tails than fibroblasts. By 24 hours of exposure to H_2O_2, repair was close to 100 % in all cell lines. In addition, untreated human ESCs had a lower level of 8-oxoguanine than fibroblasts, suggesting that either it is removed faster, or that it is accumulated at a slower rate in human ESCs. Measurement of OGG1 activity, the enzyme that removes 8-oxoguanine, showed a similar level of activity in ESCs, fibroblasts, and in the tumor cell line HeLa. This suggests that there is another enzyme that accounts for faster repair, or that there is higher antioxidant activity in ESCs that accounts for slower accumulation of 8-oxoguanine [64]. Expression profiles revealed higher expression of a number of BER genes in ESCs [62, 64] and iPSCs [62] in comparison to fibroblasts, which may explain observed high efficiency of repair of H_2O_2-induced DNA damage in ESCs.

4.3.1.4 Mismatch Repair

Mismatch repair (MMR) is responsible for repair of mismatched base pairs in DNA. Components of the MMR pathway are highly conserved and critical for the maintenance of genomic stability. Cells with defective MMR are prone to acquiring new mutations (exhibit hundreds fold higher rate of spontaneous mutations), and have a so-called "mutator" phenotype. Mismatched base pairs are most commonly produced by errors in replication. Insertion-deletion loops, formed as a result of slippage of template or primer during DNA replication, are also MMR substrates. Insertion/deletion loops may also form in heteroduplex DNA created, for example, during recombination of homologous chromosomes. The lack of MMR in cells produces microsatellite instability and a high rate of recombination events in addition to increased rate of mutations. Not surprisingly, then, in mammals, loss of MMR proteins increases susceptibility to cancer.

ESCs are rapidly dividing cells which predisposes them to increased risk of errors during replication. One of the consequences, as already described, is the risk of collapse of stalled replication forks and formation of DSBs. Another risk is incorporation of the wrong bases during successive rounds of replication, potentially leading to accumulation of mutations. These two mutational hazards suggest that ESCs need efficient mutation suppression mechanisms. Indeed, measurement of the MMR proteins MSH2, MSH6, MLHh1, and PMS2 revealed significantly elevated levels in wild type and MSH2 mutant ESCs relative to MEFs [106, 107]. Similarly, MSH2 and MSH6 mRNA were increased in ESCs [106, 107]. Higher protein expression correlated with 30-fold higher MMR activity in ESCs in *in vitro* reporter assays. In mutant MSH2$^{-/-}$ ESCs, repair efficiency was four to fivefolds lower than in wild type ESCs, but still present. Since MSH2 is involved in the earliest stages of MMR (binding to a mismatched base), mutant lines are not expected to have detectable MMR activity. Thus, the observed repair of an *in vitro* plasmid in MSH2$^{-/-}$ ESC line is carried out by another repair pathway, most likely by BER, as MMR, BER, and NER share some molecular components [106]. Another study confirmed these results and shed some light on the mechanism of methylating genotoxin N-methyl-N-nitro-N-nitrosoguanidine (MNNG) induced apoptosis in ESCs. MNNG forms multiple lesions in DNA, including O^6-methylguanine (O^6-meG). MNNG-induced O^6-meG is repaired by O^6-methylguanine DNA methyltransferase (MGMT); in the absence of MGMT, O^6-meG mispairs with thymine during replication, causing GC to AT transversion. Unrepaired O^6-meG is recognized by the MSH2/MSH6 heterodimer which then triggers apoptosis. Cells lacking either of these proteins are resistant to killing by MNNG and possess mutator phenotype. When clonogenic survival and apoptosis in response to MNNG were compared, mouse ESCs exhibited profound sensitivity relative to fibroblasts. Both cell types demonstrated similar MGMT activity, but ESCs expressed higher levels of MSH2 mRNA and protein, and exhibited stronger MSH2/MSH6 DNA binding in comparison to fibroblasts. Overexpression of MSH2 in fibroblasts resulted in increased sensitivity to MNNG, indicating that MSH2/MSH6 binding to damaged DNA was responsible for hypersensitivity to alkylating reagents. During differentiation of ESCs, expression of MSH2 and MSH6 proteins decreased followed by a concomitant decrease in MNNG-induced apoptosis. The elevated expression of MSH2 and MSH6 appeared to be due to high E2F1 activity in ESCs. In ESCs, RB is hyperphosphorylated resulting in release of active E2F1, which was found to be strongly bound to MSH2 promoter in ESCs. Finally, in ESCs treated with MNNG, H2AX was phosphorylated and p53 was localized to the nucleus, where it activated FAS receptor (Fas/CD95/Apo-1), without cytochrome c release. Thus, MNNG-induced O^6-meG can be repaired by MGMT in ESCs. Unrepaired O^6-meG is bound by MSH2/MSH6 heterodimer that signals to apoptosis via FAS receptor pathway.

In human ESCs and iPSCs, mRNA levels of multiple MMR proteins (EXO1, MLH1, MLH3, MSH2, MSH3, MUTYH, PMS1, PMS2, and N4BP2) were highly elevated relative to differentiated cells [62]. However, no studies on human ESCs and iPSCs have measured MMR activity thus far.

4.3.1.5 Genomic Integrity in Induced Pluripotent Stem Cells

Advances in iPSC technology in the past six years have created a number of interesting research and clinical opportunities. Since iPSCs are derived from somatic cells, they are spared of ethical concerns that surround ESC research, yet possess parallel developmental and differentiation potential (unlike multipotent stem cells). Derivation of iPSCs from somatic cells enables generation of patient-specific PSCs that can be differentiated into any desired cell type. Thus, iPSCs offer ethically unchallenged alternatives to ESCs for cell replacement therapies. Induced PSCs also provide tools for studying how differentiation can be reversed and how cells change their morphology, gene expression, function, and epigenetic landscape so dramatically.

Concerns about iPSCs' identity and actual similarity to ESCs were raised recently based on studies demonstrating the existence of epigenetic memory in iPSCs [108]. Furthermore, while ESCs are isolated from the ICM at the earliest developmental stage and have "pristine" genetic information, iPSCs that are derived from the adult somatic cells may have already acquired mutations during an individual's lifetime. The most commonly used method for generation of iPSCs is by transduction of somatic cells with reprogramming factors carried on integrating retro- and lenti-viruses, which causes insertional mutagenesis and could promote abnormal gene expression. Another potential issue is whether the process of reprogramming itself affects genomic integrity. Multiple studies established that repression of tumor suppressor pathways during the course of reprogramming significantly improved its efficiency [109–113]. LIN28, KLF4, and cMYC are known oncogenes that may activate tumor suppression response following their ectopic expression in somatic cells. The activation of TP53-CDKN1A pathway and *INK4/ARF* locus may induce cell cycle arrest, apoptosis, and senescence that hinder reprogramming. Furthermore, Marion et al. showed that reprogramming triggered DNA damage, which activated the DNA damage response and induced TP53-dependent apoptosis [112]. Finally, senescence of fibroblasts and increased age of a donor impaired reprogramming by upregulating the *INK4/ARF* locus. Thus, overexpression of oncogenes, integration of viral vectors into the host genome, induced DNA damage, and advanced donor age may activate TP53-CDKN1A or INK4/ARF that act as a roadblock during reprogramming by inducing cell cycle arrest, apoptosis, and senescence [109–113].

Studies in multipotent stem cells have revealed that TP53 plays a role in regulating cellular differentiation. For example, DSBs induced TP53 acetylation at lysine 320 in the central nervous system (CNS) [114], which was required for the promotion of neurite outgrowth in the cell culture and axonal regeneration in mice [115]. Loss of TP53 in neural stem cells (NSCs) in the subventricular zone resulted in increased NSC proliferation and self-renewal *in vivo* [116, 117]. Concomitant deletion of TP53 and PTEN in mouse CNS promoted an undifferentiated state with high self-renewal potential, and impaired expression of glial and neuronal lineage markers [118]. Similarly, both physiological and extrinsically-induced DSBs in developing B cells signal through ATM to induce B cell specific gene expression. Inhibition of ATM signaling promoted B-cell proliferation and caused differentiation failure [119]. Correspondingly, TP53 was shown to negatively regulate self-renewal and promote quiescence in hematopoietic stem cells (HSCs) [120]. Collectively, these studies demonstrate the role of TP53 in negatively regulating self-renewal and promoting differentiation of stem cells in multiple tissues, including CNS and hematopoietic systems. Therefore, loss of TP53 and *INK4/ARF* may facilitate somatic cell reprogramming by promoting self-renewal [121]. In this respect, the reprogramming process resembles malignant transformation, and loss of TP53 may promote formation of cancer stem cells with the ability of self-renewal. It is very unlikely that iPSCs generated by permanent knock-down of TP53 will be used in clinical applications due to the increased risk of genomic instability and malignant transformation [121]. Indeed, when mouse iPSCs generated from terminally differentiated TP53 deficient T cells were injected into mouse blastocyst, they contributed to the development of chimeric mice, but these mice developed tumors and died within seven weeks [109].

These findings raise further questions—what is the status of TP53 in iPSCs, and how do iPSCs respond to DNA damage? While the studies outlined above clearly demonstrate the role of TP53 and INK4/ARF in the suppression of reprogramming, it remains to be answered whether only cells that have spontaneously mutated or epigenetically silenced TP53 or INK4/ARF tumor suppressors can be reprogrammed, or if the activation of TP53 and INK4/ARF prevents reprogramming of cells with damaged DNA in order to avoid generation of PSCs with defective DNA repair and genomic instability [122]. Thus far, few studies have addressed these questions. Our results demonstrated that human iPSCs were extremely radiosensitive and underwent extensive apoptosis within hours of exposure to ionizing radiation. Ionizing radiation treatment induced ATM, CHEK2, and TP53 phosphorylation and their proper localization in human iPSCs [62]. Furthermore, irradiated iPSCs arrested in the G_2/M phase of the cell cycle and repaired DSBs as efficiently as human ESCs. Expression of a number of DNA repair genes and proteins was elevated in both iPSCs and ESCs in comparison to their differentiated counterparts [62, 97]. When the accuracy of NHEJ in human ESCs, iPSCs, and their parent cells was compared, ESCs and iPSCs showed accurate NHEJ, whereas more than 20 nucleotides were frequently deleted in parental cells [97]. Thus, the DDR in human iPSCs strongly resembles the one in human ESCs, and indicates that during reprogramming, the DDR of somatic cells undergoes dramatic changes.

The genomic integrity of iPSCs has been under investigation recently. Several studies noticed the presence of large chromosomal rearrangements and copy number variations (CNVs) in reprogrammed cells that were largely absent in parent fibroblast lines [123–125]. Sequencing of protein coding genomic regions (exome) of 22 human iPSC lines and nine matched fibroblast lines identified known and novel variants for each line, whereas capillary sequencing subsequently validated 124 mutations. The observed mutational load was tenfold higher in iPSCs than normal cell culture mutational load. Importantly, most of the iPSC lines derived from the same fibroblast line did not share common mutations [126]. Thus, mutations in iPSCs may (1) pre-exist in fibroblast population at low frequencies, (2) arise during the reprogramming process, or (3) arise after reprogramming, during the clone isolation and expansion. The data suggest that some of the mutations existed in parental fibroblasts, although at very low frequencies, while the others occurred during reprogramming and subsequent culturing [126]. For example, large numbers of novel CNVs were detected in early passages of iPSCs, but subsequent *in vitro* culturing selected against most of these changes, indicating that they were deleterious [123]. In iPSC lines generated by expression of Simian virus 40 large T antigen (a TP53 suppressor) or from RB1 mutant fibroblasts, mutational burden was similar to other iPSC lines, indicating that tumor suppression cannot account for the total of reprogramming associated mutations. However, the possibility that reprogramming is a mutagenic process *per se* has not been completely excluded [126]. Finally, it appears that during the culturing of iPSCs, novel mutations may arise as a result of selection of cells with the growth advantages. Indeed, genes carrying mutations in iPSCs were enriched for genes mutated in cancers [123, 125, 126]. Nevertheless, the functional significance of the CNVs and point mutations is still largely unknown since many of them are not associated with known disease-related genetic markers [127]. In contrast to these findings, Quinlan and colleagues reported retroelement stability and rare DNA rearrangements during MEF reprogramming [128]. *De novo* structural variants (SVs; insertions, deletions, duplications, inversions, translocations, and transposition of endogenous retroelements) were detected using whole genome paired-end DNA sequencing and sensitive SV detection algorithms. Surprisingly, only four variants were found in three iPSC lines. The observed discrepancy in results between studies on the frequency of *de novo* CNVs, SVs, and point mutations in human [123–126] and mouse [128] iPSCs could reflect either species-specific differences or the difference in reprogramming method (using adult versus embryonic somatic cells) and study design (comparing early versus late passage iPSCs, whole genome sequencing versus microarray based approaches, different algorithms for data analysis).

Thus, before iPSCs can be used in any clinical applications, it will be essential to reveal which aspects of reprogramming and subsequent culturing might induce the observed genetic and epigenetic abnormalities in human iPSCs and to understand the biological significance of these changes. This includes their effects on self-renewal and proliferation, differentiation, tumorigenicity and functionality of iPSCs and their progeny [127].

4.3.2 DNA Damage Responses in Multipotent Stem Cells

Multipotent stem cells are long-lived and function to replace damaged or dead cells over an organism's life. Their capacity to repair DNA during the lifetime diminishes and old cells contain more DNA lesions than young cells. This is particularly exacerbated in patients with premature aging syndromes, such as Hutchinson-Gilford progeria, Werner's syndrome, or Cockayne's syndrome. One underlying characteristic of these syndromes is a greater than normal accumulation of DNA damage. With the exception of Hutchinson-Gilford progeria, these syndromes are caused by mutations in DNA repair genes, clearly demonstrating the role of DNA repair in longevity. Since multipotent stem cells are long-lived, they are exposed to a lifetime of accumulation of DNA lesions. Accurate DNA repair is essential for maintenance of stem cell function and tissue homeostasis. Unrepaired DNA damage may induce apoptosis or senescence leading to depletion of stem cell reserves and aging. However, if adult stem cells acquire resistance to genotoxic stressors, they may evade apoptosis and acquire mutations that promote tumorigenesis. Defective DNA repair can promote accumulation of mutations that interfere with self-renewal and differentiation, or induce malignant transformation (Fig. 4.5). Therefore, there is a keen interest in mechanisms that safeguard genomic integrity of adult stem cells, and in particular the function of DNA repair mechanisms.

Most adult stem cells are quiescent (but not all—intestine stem cells are actively cycling), and spend most of their life in the G_0 phase of the cell cycle. It has been postulated that quiescence preserves the genomic integrity and function of stem cells by limiting the risk of replication errors [129]. In addition, most of the adult stem cells have low metabolic activity and low production of ROS. This is mainly

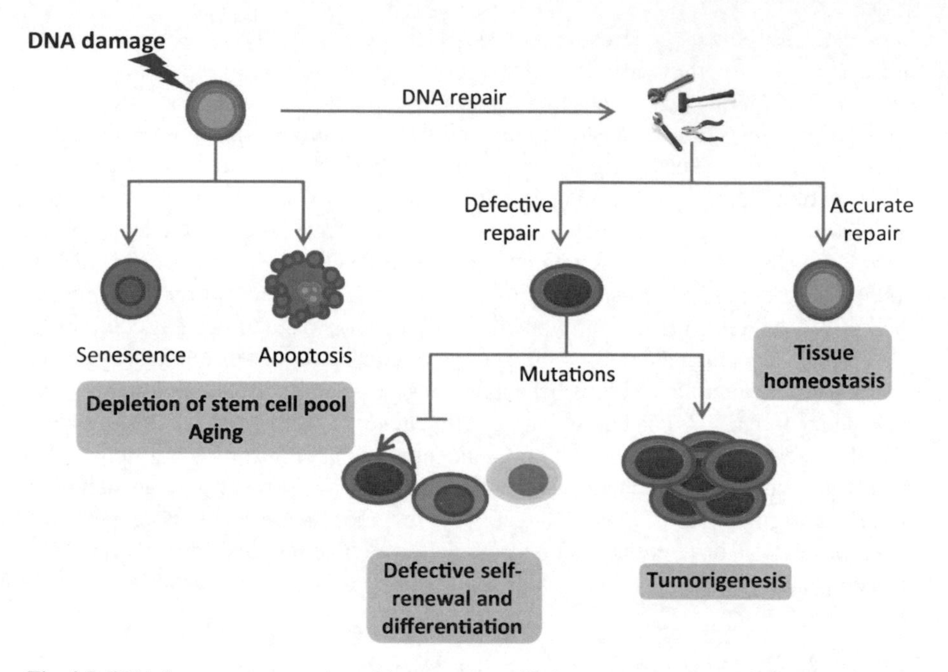

Fig. 4.5 DNA damage responses in multipotent stem cells. In response to genotoxic stress, quiescent multipotent stem cells may undergo senescence or apoptosis. This would maintain a pristine cellular population, but might cause depletion of the stem cell reserves and have adverse effect on tissue homeostasis. Accurate DNA repair protects damaged multipotent stem cells from accumulation of mutations and contributes to their survival and function. However, quiescence of adult stem cells limits their choice of double strand break repair pathway, and cells may have to rely on inaccurate repair. Defective DNA repair would allow for accumulation of mutations that may interfere with the self-renewal and differentiation, or may promote malignant transformation

because adult stem cells, similar to ESCs, rely on the glycolytic pathway to produce ATP, whereas differentiated somatic cells depend on mitochondria to meet their energy demands. Mitochondrial respiration is a major source of cellular ROS, and by relying on glycolysis for energy production, stem cells minimize production of ROS and oxidative stress [130–132]. Stem cells also have high ABC transporter activity which provides efficient efflux of genotoxins. Quiescence, however, may have adverse effects on the genomic integrity of adult stem cells, because checkpoint function and some repair pathways are cell cycle dependent [75, 79]. Stem cells that reside in the G_0 phase of the cell cycle have NHEJ as their only option for repair of DSBs. As we already discussed, NHEJ is regarded as an error prone repair pathway, and may lead to acquisition of new mutations in stem cells [79]. Indeed, irradiation of quiescent mouse HSCs leads to formation of chromosomal rearrangements, whereas irradiation of HSCs which are driven into the cell cycle results in lower frequency of genomic rearrangements due to utilization of high fidelity HR [133]. Nevertheless, NHEJ has an essential role in DNA repair in HSCs, and lack of XRCC4 or LIG4 is associated with the reduced repopulation potential of HSCs [134, 135].

Quiescence may contribute to resistance to apoptosis following irradiation [136]. Tissues with high turnover, such as intestine and embryonic tissue, are very sensitive to radiation-induced DNA damage [137, 138], whereas many quiescent tissue-specific stem cells and cancer stem cells are more resistant to killing by irradiation [139, 140]. For example, mouse HSCs are quiescent and more resistant to irradiation in comparison to downstream progenitors, but the short-term radioresistance and rapid DNA repair by error-prone NHEJ abrogate repopulation potential of irradiated HSCs in serial transplantation assay [133]. Studies on HSCs with defects in DNA repair pathways revealed that mutant HSCs have decreased ability to reconstitute bone marrow in competitive transplantation assays, which becomes more prominent with the increasing age of the donor and elapsed time post transplantation (long-term reconstitution was most affected). Interestingly, the endogenous DNA damage was replication independent, since genomes of quiescent HSCs contained more γH2AX foci than genomes of proliferating progenitor cells. There are two possible explanations: (1) progenitors are more likely to have a sister chromatid available in comparison to quiescent HSCs and can utilize HR for DNA repair; (2) progenitors are more prone to apoptosis than HSCs. Thus, HSCs are quiescent cells, relatively resistant to apoptosis and rely on error-prone DNA repair; longevity and error-prone DNA repair promote accumulation of DNA changes in HSCs. Conversely, proliferating progenitors are more prone to apoptosis leading to loss of cells with DNA damage. The net result is enrichment of bone marrow with HSCs that have accumulated DNA mutations and have impaired self-renewal and proliferation, and loss of progenitors that give rise to differentiated cells, ultimately leading to hypocellularity and aging, or malignant transformation [134, 135, 141].

Interestingly, human HSCs derived from umbilical cord blood are more radiosensitive than downstream progenitor cells. Human HSCs also show slower kinetics of DSB repair. Irradiation abrogated human HSCs potential to repopulate bone marrow of recipient NOD/SCID mice. Knock-down of TP53, or overexpression of prosurvival factor BCL2 rescued repopulation capacity of irradiated human HSCs in the short term, but irradiated HSCs lacking TP53 were unable to sustain hematopoiesis and contained persistent DSBs in secondary recipients [142]. Increased expression of BCL2, reliance on NHEJ for DSB repair, and resistance to irradiation have also been reported in hair follicle bulge stem cells of mouse epidermis [143].

The differential sensitivities of adult stem cells to DNA damage-induced apoptosis may modulate cancer predisposition of a particular tissue. For example, in high turnover tissues with low apoptotic threshold such as intestine, damaged cells are efficiently eliminated thereby shielding intestine from accumulation of cells with unrepaired DNA damage and cancer development [144]. In contrast, colon stem cells are more resistant to apoptosis in the presence of DNA damage, which may explain increased frequency of colon cancers relative to intestine [145]. However, a lower apoptotic threshold may adversely affect the regenerative potential of an organ due to depletion of stem cell reserves.

Oxidative stress and ROS can introduce significant DNA damage burden, and greatly affect stem cell genomic stability, viability, and function. For example, FOXO proteins regulate expression of anti-ROS defense enzymes and reduce ROS levels.

Mice that lack all three FOXO proteins show depletion of HSCs and NSCs [146–148]. HSCs of FOXO null mice exhibit poor transplantation potential, whereas NSCs undergo poor proliferation and differentiation. Similarly, loss of ATM causes progressive bone marrow failure in adult mice due to increased ROS production [149, 150]. Treatment of ATM-null and FOXO-null mice with antioxidants can reverse stem cell defects, clearly demonstrating the role of increased ROS concentration in stem cell compromise [147, 149]. Another important stem cell regulator, BMI1, is involved in regulation of mitochondrial function: deletion of BMI1 is associated with decreased mitochondrial function and increased ROS production, which adversely affect stem cells [151, 152]. BMI1 overexpression can enhance ATM recruitment to the sites of DSBs in human NSCs after UV treatment [153]. Thus, BMI1 plays dual role in stem cells by negatively regulating ROS levels and by promoting ATM localization to DSBs [152, 153]. ROS are also important signaling molecules that can modulate lineage commitment, and increased ROS levels may promote differentiation of HSCs [154].

4.4 Conclusions

Stem cells have remarkable capabilities to undergo self-renewal and differentiation. PSCs, such as ESCs and iPSCs, can undergo self-renewal indefinitely in cell culture, and yet are able to form any type of somatic cell. Multipotent stem cells are adult stem cells that form a limited number of related cell types, but are present in tissues and organs throughout an organism's lifespan. While PSCs exist only short-term during development, multipotent stem cells are long-lived and are responsible for the replacement of damaged and dead somatic cells. Therefore, any genetic alterations in stem cells would be passaged to the daughter cells and could affect one or more cell lineages. It is therefore essential, for stem cells to safeguard their genomes and prevent acquisition of DNA mutations. In PSCs this is achieved in part by minimizing induction of DNA lesions by maintaining low levels of ROS, high activity of the efflux pumps, expression of telomerase, and by efficient DNA repair. Both mouse and human ESCs exhibit elevated expression of DNA repair proteins and proficiently repair DNA lesions. Mouse ESCs mostly depend on HR for accurate DSB repair, whereas human ESCs utilize both HR and NHEJ for repair of DSBs. Preservation of pristine populations of ESCs is also facilitated by a low threshold for apoptosis.

Induced PSCs closely resemble ESCs in many aspects, but we are only beginning to elucidate similarities and differences between iPSCs and ESCs. An important distinction between these two PSC types is that ESCs are derived from the blastocyst stage embryos, whereas iPSCs are produced from adult somatic cells that could have accumulated genetic changes during the lifetime. Therefore, it is important to elucidate the genomic integrity and mutational burden in iPSCs. Available data suggest that iPSCs acquire ESC-like responses to DNA damage and exhibit extreme radiosensitivity, elevated expression of DNA repair proteins, and efficient DNA repair. Nevertheless, numerous challenges in understanding the nature of iPSCs and the impact of reprogramming on their genomic stability remain.

Multipotent stem cells differ significantly from ESCs. Unlike ESCs, most of the multipotent stem cells are slow dividing quiescent cells. A quiescent state allows multipotent stem cells to maintain low levels of metabolic activity and ROS production, thereby minimizing cellular oxidative stress. In addition, multipotent stem cells have high activity of ABC transporters, aiding in efflux of potentially genotoxic chemicals. Furthermore, quiescence reduces the risk of replicative errors, which are particularly dangerous in rapidly proliferating ESCs. Although mostly protective, quiescence of multipotent stem cells has an adverse effect on DNA repair—while they are resting in the G_0 phase of the cell cycle, multipotent stem cells cannot perform HR, and have to rely on error-prone NHEJ for DSB repair. Long life and error-prone DNA repair may predispose multipotent stem cells to acquisition of mutations that interfere with their normal function, and may lead to aging, disease, or malignant transformation.

Acknowledgments This research was supported by grants from the National Institute of Child Health and Development, 1PO1HD047675, and California Institute of Regenerative Medicine, TG2-01155. For providing critical assistance we thank Xianmin Zeng.

References

1. Friedberg EC, Walker GC, Siede W, Wood RD, Schultz RA, Ellenberger T (2006) DNA repair and mutagenesis. ASM Press, Washington
2. Kahan BW, Ephrussi B (1970) Developmental potentialities of clonal in vitro cultures of mouse testicular teratoma. J Natl Cancer Inst 44:1015–1036
3. Kleinsmith LJ, Pierce GB Jr (1964) Multipotentiality of single embryonal carcinoma cells. Cancer Res 24:1544–1551
4. Atkin NB, Baker MC, Robinson R, Gaze SE (1974) Chromosome studies on 14 near-diploid carcinomas of the ovary. Eur J Cancer 10:144–146
5. Matsui Y, Zsebo K, Hogan BL (1992) Derivation of pluripotential embryonic stem cells from murine primordial germ cells in culture. Cell 70:841–847
6. Resnick JL, Bixler LS, Cheng L, Donovan PJ (1992) Long-term proliferation of mouse primordial germ cells in culture. Nature 359:550–551
7. Stewart CL, Gadi I, Bhatt H (1994) Stem cells from primordial germ cells can reenter the germ line. Dev Biol 161:626–628
8. Labosky PA, Barlow DP, Hogan BL (1994) Mouse embryonic germ (EG) cell lines: transmission through the germline and differences in the methylation imprint of insulin-like growth factor 2 receptor (Igf2r) gene compared with embryonic stem (ES) cell lines. Development 120:3197–3204
9. Tada M, Tada T, Lefebvre L, Barton SC, Surani MA (1997) Embryonic germ cells induce epigenetic reprogramming of somatic nucleus in hybrid cells. Embo J 16:6510–6520
10. Shamblott MJ, Axelman J, Wang S, Bugg EM, Littlefield JW, Donovan PJ, Blumenthal PD, Huggins GR, Gearhart JD (1998) Derivation of pluripotent stem cells from cultured human primordial germ cells. Proc Natl Acad Sci USA 95:13726–13731
11. Tesar PJ, Chenoweth JG, Brook FA, Davies TJ, Evans EP, Mack DL, Gardner RL, McKay RD (2007) New cell lines from mouse epiblast share defining features with human embryonic stem cells. Nature 448:196–199
12. Brons IG, Smithers LE, Trotter MW, Rugg-Gunn P, Sun B, Chuva de Sousa Lopes SM, Howlett SK, Clarkson A, Ahrlund-Richter L, Pedersen RA, Vallier L (2007) Derivation of pluripotent epiblast stem cells from mammalian embryos. Nature 448:191–195

13. Hogan B, Fellous M, Avner P, Jacob F (1977) Isolation of a human teratoma cell line which expresses F9 antigen. Nature 270:515–518
14. Martin GR (1981) Isolation of a pluripotent cell line from early mouse embryos cultured in medium conditioned by teratocarcinoma stem cells. Proc Natl Acad Sci USA 78:7634–7638
15. Evans MJ, Kaufman MH (1981) Establishment in culture of pluripotential cells from mouse embryos. Nature 292:154–156
16. Thomson JA, Itskovitz-Eldor J, Shapiro SS, Waknitz MA, Swiergiel JJ, Marshall VS, Jones JM (1998) Embryonic stem cell lines derived from human blastocysts. Science 282:1145–1147
17. Takahashi K, Yamanaka S (2006) Induction of pluripotent stem cells from mouse embryonic and adult fibroblast cultures by defined factors. Cell 126:663–676
18. Takahashi K, Tanabe K, Ohnuki M, Narita M, Ichisaka T, Tomoda K, Yamanaka S (2007) Induction of pluripotent stem cells from adult human fibroblasts by defined factors. Cell 131:861–872
19. Tam PP, Rossant J (2003) Mouse embryonic chimeras: tools for studying mammalian development. Development 130:6155–6163
20. Kubiak JZ, Tarkowski AK (1985) Electrofusion of mouse blastomeres. Exp Cell Res 157:561–566
21. Nagy A, Gocza E, Diaz EM, Prideaux VR, Ivanyi E, Markkula M, Rossant J (1990) Embryonic stem cells alone are able to support fetal development in the mouse. Development 110:815–821
22. Nagy A, Rossant J, Nagy R, Abramow-Newerly W, Roder JC (1993) Derivation of completely cell culture-derived mice from early-passage embryonic stem cells. Proc Natl Acad Sci USA 90:8424–8428
23. Nichols J, Smith A (2009) Naive and primed pluripotent states. Cell Stem Cell 4:487–492
24. Hanna J, Cheng AW, Saha K, Kim J, Lengner CJ, Soldner F, Cassady JP, Muffat J, Carey BW, Jaenisch R (2010) Human embryonic stem cells with biological and epigenetic characteristics similar to those of mouse ESCs. Proc Natl Acad Sci USA 107:9222–9227
25. Boyer LA, Lee TI, Cole MF, Johnstone SE, Levine SS, Zucker JP, Guenther MG, Kumar RM, Murray HL, Jenner RG, Gifford DK, Melton DA, Jaenisch R, Young RA (2005) Core transcriptional regulatory circuitry in human embryonic stem cells. Cell 122:947–956
26. Chambers I, Colby D, Robertson M, Nichols J, Lee S, Tweedie S, Smith A (2003) Functional expression cloning of Nanog, a pluripotency sustaining factor in embryonic stem cells. Cell 113:643–655
27. Mitsui K, Tokuzawa Y, Itoh H, Segawa K, Murakami M, Takahashi K, Maruyama M, Maeda M, Yamanaka S (2003) The homeoprotein Nanog is required for maintenance of pluripotency in mouse epiblast and ES cells. Cell 113:631–642
28. Niwa H, Miyazaki J, Smith AG (2000) Quantitative expression of Oct-3/4 defines differentiation, dedifferentiation or self-renewal of ES cells. Nat Genet 24:372–376
29. Okamoto K, Okazawa H, Okuda A, Sakai M, Muramatsu M, Hamada H (1990) A novel octamer binding transcription factor is differentially expressed in mouse embryonic cells. Cell 60:461–472
30. Davis RL, Weintraub H, Lassar AB (1987) Expression of a single transfected cDNA converts fibroblasts to myoblasts. Cell 51:987–1000
31. Niwa H, Toyooka Y, Shimosato D, Strumpf D, Takahashi K, Yagi R, Rossant J (2005) Interaction between Oct3/4 and Cdx2 determines trophectoderm differentiation. Cell 123:917–929
32. Xie H, Ye M, Feng R, Graf T (2004) Stepwise reprogramming of B cells into macrophages. Cell 117:663–676
33. Campbell KH, McWhir J, Ritchie WA, Wilmut I (1996) Sheep cloned by nuclear transfer from a cultured cell line. Nature 380:64–66
34. Yu J, Vodyanik MA, He P, Slukvin II, Thomson JA (2006) Human embryonic stem cells reprogram myeloid precursors following cell-cell fusion. Stem Cells 24:168–176
35. Yu J, Vodyanik MA, Smuga-Otto K, Antosiewicz-Bourget J, Frane JL, Tian S, Nie J, Jonsdottir GA, Ruotti V, Stewart R, Slukvin II, Thomson JA (2007) Induced pluripotent stem cell lines derived from human somatic cells. Science 318:1917–1920

36. Yu J, Hu K, Smuga-Otto K, Tian S, Stewart R, Slukvin II, Thomson JA (2009) Human induced pluripotent stem cells free of vector and transgene sequences. Science 324:797–801
37. Nakagawa M, Koyanagi M, Tanabe K, Takahashi K, Ichisaka T, Aoi T, Okita K, Mochiduki Y, Takizawa N, Yamanaka S (2008) Generation of induced pluripotent stem cells without Myc from mouse and human fibroblasts. Nat Biotechnol 26:101–106
38. Park IH, Arora N, Huo H, Maherali N, Ahfeldt T, Shimamura A, Lensch MW, Cowan C, Hochedlinger K, Daley GQ (2008) Disease-specific induced pluripotent stem cells. Cell 134:877–886
39. Kim JB, Zaehres H, Wu G, Gentile L, Ko K, Sebastiano V, Arauzo-Bravo MJ, Ruau D, Han DW, Zenke M, Scholer HR (2008) Pluripotent stem cells induced from adult neural stem cells by reprogramming with two factors. Nature 454:646–650
40. Kim D, Kim CH, Moon JI, Chung YG, Chang MY, Han BS, Ko S, Yang E, Cha KY, Lanza R, Kim KS (2009) Generation of human induced pluripotent stem cells by direct delivery of reprogramming proteins. Cell Stem Cell 4:472–476
41. Fusaki N, Ban H, Nishiyama A, Saeki K, Hasegawa M (2009) Efficient induction of transgene-free human pluripotent stem cells using a vector based on Sendai virus, an RNA virus that does not integrate into the host genome. Proc Jpn Acad Ser B Phys Biol Sci 85:348–362
42. Friedrich Ben-Nun I, Montague SC, Houck ML, Tran HT, Garitaonandia I, Leonardo TR, Wang YC, Charter SJ, Laurent LC, Ryder OA, Loring JF (2011) Induced pluripotent stem cells from highly endangered species. Nat Methods 8:829–831
43. Sahin E, Depinho RA (2010) Linking functional decline of telomeres, mitochondria and stem cells during ageing. Nature 464:520–528
44. Knoblich JA (2001) Asymmetric cell division during animal development. Nat Rev Mol Cell Biol 2:11–20
45. Lajtha LG (1979) Stem cell concepts. Nouv Rev Fr Hematol 21:59–65
46. Clarke MF, Fuller M (2006) Stem cells and cancer: two faces of eve. Cell 124:1111–1115
47. Savatier P, Huang S, Szekely L, Wiman KG, Samarut J (1994) Contrasting patterns of retinoblastoma protein expression in mouse embryonic stem cells and embryonic fibroblasts. Oncogene 9:809–818
48. Savatier P, Lapillonne H, van Grunsven LA, Rudkin BB, Samarut J (1996) Withdrawal of differentiation inhibitory activity/leukemia inhibitory factor up-regulates D-type cyclins and cyclin-dependent kinase inhibitors in mouse embryonic stem cells. Oncogene 12:309–322
49. White J, Stead E, Faast R, Conn S, Cartwright P, Dalton S (2005) Developmental activation of the Rb-E2F pathway and establishment of cell cycle-regulated cyclin-dependent kinase activity during embryonic stem cell differentiation. Mol Biol Cell 16:2018–2027
50. Becker KA, Ghule PN, Therrien JA, Lian JB, Stein JL, van Wijnen AJ, Stein GS (2006) Self-renewal of human embryonic stem cells is supported by a shortened G1 cell cycle phase. J Cell Physiol 209:883–893
51. Becker KA, Stein JL, Lian JB, van Wijnen AJ, Stein GS (2010) Human embryonic stem cells are pre-mitotically committed to self-renewal and acquire a lengthened G1 phase upon lineage programming. J Cell Physiol 222:103–110
52. Stead E, White J, Faast R, Conn S, Goldstone S, Rathjen J, Dhingra U, Rathjen P, Walker D, Dalton S (2002) Pluripotent cell division cycles are driven by ectopic Cdk2, cyclin A/E and E2F activities. Oncogene 21:8320–8333
53. Momcilovic O, Navara C, Schatten G (2011) Cell cycle adaptations and maintenance of genomic integrity in embryonic stem cells and induced pluripotent stem cells. Results Probl Cell Differ 53:415–458
54. Momcilovic O, Choi S, Varum S, Bakkenist C, Schatten G, Navara C (2009) Ionizing radiation induces ataxia telangiectasia mutated-dependent checkpoint signaling and G(2) but not G(1) cell cycle arrest in pluripotent human embryonic stem cells. Stem Cells 27:1822–1835
55. Hong Y, Stambrook PJ (2004) Restoration of an absent G1 arrest and protection from apoptosis in embryonic stem cells after ionizing radiation. Proc Natl Acad Sci USA 101:14443–14448
56. Fluckiger AC, Marcy G, Marchand M, Negre D, Cosset FL, Mitalipov S, Wolf D, Savatier P, Dehay C (2006) Cell cycle features of primate embryonic stem cells. Stem Cells 24:547–556

57. Filion TM, Qiao M, Ghule PN, Mandeville M, van Wijnen A J, Stein JL, Lian JB, Altieri DC, Stein G.S (2009) Survival responses of human embryonic stem cells to DNA damage. J Cell Physiol 220:586–592
58. Barta T, Vinarsky V, Holubcova Z, Dolezalova D, Verner J, Pospisilova S, Dvorak P, Hampl A (2010) Human embryonic stem cells are capable of executing G1/S checkpoint activation. Stem Cells 28:1143–1152
59. Lin T, Chao C, Saito S, Mazur SJ, Murphy ME, Appella E, Xu Y (2005) p53 induces differentiation of mouse embryonic stem cells by suppressing Nanog expression. Nat Cell Biol 7:165–171
60. Qin H, Yu T, Qing T, Liu Y, Zhao Y, Cai J, Li J, Song Z, Qu X, Zhou P, Wu J, Ding M, Deng H (2007) Regulation of apoptosis and differentiation by p53 in human embryonic stem cells. J Biol Chem 282:5842–5852
61. Wilson KD, Sun N, Huang M, Zhang WY, Lee AS, Li Z, Wang SX, Wu JC (2010) Effects of ionizing radiation on self-renewal and pluripotency of human embryonic stem cells. Cancer Res 70:5539–5548
62. Momcilovic O, Knobloch L, Fornsaglio J, Varum S, Easley C, Schatten G (2010) DNA damage responses in human induced pluripotent stem cells and embryonic stem cells. PLoS One 5:e13410
63. Saretzki G, Armstrong L, Leake A, Lako M, von Zglinicki T (2004) Stress defense in murine embryonic stem cells is superior to that of various differentiated murine cells. Stem Cells 22:962–971
64. Maynard S, Swistikowa AM, Lee JW, Liu Y, Liu ST, A DAC, Rao M, de Souza-Pinto N, Zeng X, Bohr VA (2008) Human embryonic stem cells have enhanced repair of multiple forms of DNA damage. Stem Cells 26:2266–2274
65. Saretzki G, Walter T, Atkinson S, Passos JF, Bareth B, Keith WN, Stewart R, Hoare S, Stojkovic M, Armstrong L, von Zglinicki T, Lako M (2008) Downregulation of multiple stress defense mechanisms during differentiation of human embryonic stem cells. Stem Cells 26:455–464
66. Rich T, Allen RL, Wyllie AH (2000) Defying death after DNA damage. Nature 407:777–783
67. Haber JE (2000) Partners and pathwaysrepairing a double-strand break. Trends Genet 16:259–264
68. Rothkamm K, Kruger I, Thompson LH, Lobrich M (2003) Pathways of DNA double-strand break repair during the mammalian cell cycle. Mol Cell Biol 23:5706–5715
69. Lees-Miller SP, Meek K (2003) Repair of DNA double strand breaks by non-homologous end joining. Biochimie 85:1161–1173
70. Lieber MR, Ma Y, Pannicke U, Schwarz K (2003) Mechanism and regulation of human non-homologous DNA end-joining. Nat Rev Mol Cell Biol 4:712–720
71. Mahaney BL, Meek K, Lees-Miller SP (2009) Repair of ionizing radiation-induced DNA double-strand breaks by non-homologous end-joining. Biochem J 417:639–650
72. Lee SE, Mitchell RA, Cheng A, Hendrickson EA (1997) Evidence for DNA-PK-dependent and -independent DNA double-strand break repair pathways in mammalian cells as a function of the cell cycle. Mol Cell Biol 17:1425–1433
73. Takata M, Sasaki MS, Sonoda E, Morrison C, Hashimoto M, Utsumi H, Yamaguchi-Iwai Y, Shinohara A, Takeda S (1998) Homologous recombination and non-homologous end-joining pathways of DNA double-strand break repair have overlapping roles in the maintenance of chromosomal integrity in vertebrate cells. Embo J 17:5497–5508
74. Pierce AJ, Hu P, Han M, Ellis N, Jasin M (2001) Ku DNA end-binding protein modulates homologous repair of double-strand breaks in mammalian cells. Genes Dev 15:3237–3242
75. Kim JS, Krasieva TB, Kurumizaka H, Chen DJ, Taylor AM, Yokomori K (2005) Independent and sequential recruitment of NHEJ and HR factors to DNA damage sites in mammalian cells. J Cell Biol 170:341–347
76. Lundin C, Schultz N, Arnaudeau C, Mohindra A, Hansen LT, Helleday T (2003) RAD51 is involved in repair of damage associated with DNA replication in mammalian cells. J Mol Biol 328:521–535

77. Shrivastav M, Miller CA, De Haro LP, Durant ST, Chen BP, Chen DJ, Nickoloff JA (2009) DNA-PKcs and ATM co-regulate DNA double-strand break repair. DNA Repair (Amst) 8:920–929
78. Mao Z, Bozzella M, Seluanov A, Gorbunova V (2008) DNA repair by nonhomologous end joining and homologous recombination during cell cycle in human cells. Cell Cycle 7:2902–2906
79. Mao Z, Bozzella M, Seluanov A, Gorbunova V (2008) Comparison of nonhomologous end joining and homologous recombination in human cells. DNA Repair (Amst) 7:1765–1771
80. Tsuzuki T, Fujii Y, Sakumi K, Tominaga Y, Nakao K, Sekiguchi M, Matsushiro A, Yoshimura Y, Morita T (1996) Targeted disruption of the Rad51 gene leads to lethality in embryonic mice. Proc Natl Acad Sci USA 93:6236–6240
81. Lim DS, Hasty P (1996) A mutation in mouse rad51 results in an early embryonic lethal that is suppressed by a mutation in p53. Mol Cell Biol 16:7133–7143
82. Xiao Y, Weaver DT (1997) Conditional gene targeted deletion by Cre recombinase demonstrates the requirement for the double-strand break repair Mre11 protein in murine embryonic stem cells. Nucl Acids Res 25:2985–2991
83. Zhu J, Petersen S, Tessarollo L, Nussenzweig A (2001) Targeted disruption of the Nijmegen breakage syndrome gene NBS1 leads to early embryonic lethality in mice. Curr Biol 11:105–109
84. Luo G, Yao MS, Bender CF, Mills M, Bladl AR, Bradley A, Petrini JH (1999) Disruption of mRad50 causes embryonic stem cell lethality, abnormal embryonic development, and sensitivity to ionizing radiation. Proc Natl Acad Sci USA 96:7376–7381
85. Essers J, Hendriks RW, Swagemakers SM, Troelstra C, de Wit J, Bootsma D, Hoeijmakers JH, Kanaar R (1997) Disruption of mouse RAD54 reduces ionizing radiation resistance and homologous recombination. Cell 89:195–204
86. Rijkers T, Van Den Ouweland J, Morolli B, Rolink AG, Baarends WM, Van Sloun PP, Lohman PH, Pastink A (1998) Targeted inactivation of mouse RAD52 reduces homologous recombination but not resistance to ionizing radiation. Mol Cell Biol 18:6423–6429
87. Barnes DE, Stamp G, Rosewell I, Denzel A, Lindahl T (1998) Targeted disruption of the gene encoding DNA ligase IV leads to lethality in embryonic mice. Curr Biol 8:1395–1398
88. Frank KM, Sharpless NE, Gao Y, Sekiguchi JM, Ferguson DO, Zhu C, Manis JP, Horner J, DePinho RA, Alt FW (2000) DNA ligase IV deficiency in mice leads to defective neurogenesis and embryonic lethality via the p53 pathway. Mol Cell 5:993–1002
89. Li Z, Otevrel T, Gao Y, Cheng HL, Seed B, Stamato TD, Taccioli GE, Alt FW (1995) The XRCC4 gene encodes a novel protein involved in DNA double-strand break repair and V(D)J recombination. Cell 83:1079–1089
90. Gao Y, Chaudhuri J, Zhu C, Davidson L, Weaver DT, Alt FW (1998) A targeted DNA-PKcs-null mutation reveals DNA-PK-independent functions for KU in V(D)J recombination. Immunity 9:367–376
91. Gu Y, Jin S, Gao Y, Weaver DT, Alt FW (1997) Ku70-deficient embryonic stem cells have increased ionizing radiosensitivity, defective DNA end-binding activity, and inability to support V(D)J recombination. Proc Natl Acad Sci USA 94:8076–8081
92. Jin S, Inoue S, Weaver DT (1998) Differential etoposide sensitivity of cells deficient in the Ku and DNA-PKcs components of the DNA-dependent protein kinase. Carcinogenesis 19:965–971
93. Nussenzweig A, Sokol K, Burgman P, Li L, Li GC (1997) Hypersensitivity of Ku80-deficient cell lines and mice to DNA damage: the effects of ionizing radiation on growth, survival, and development. Proc Natl Acad Sci USA 94:13588–13593
94. Serrano L, Liang L, Chang Y, Deng L, Maulion C, Nguyen S, Tischfield JA (2011) Homologous recombination conserves DNA sequence integrity throughout the cell cycle in embryonic stem cells. Stem Cells Dev 20:363–374
95. Tichy ED, Pillai R, Deng L, Liang L, Tischfield J, Schwemberger SJ, Babcock GF, Stambrook PJ (2010) Mouse embryonic stem cells, but not somatic cells, predominantly use homologous recombination to repair double-strand DNA breaks. Stem Cells Dev 19:1699–1711

96. Banuelos CA, Banath JP, Macphail SH, Zhao J, Eaves CA, O'Connor MD, Lansdorp PM, Olive PL (2008) Mouse but not human embryonic stem cells are deficient in rejoining of ionizing radiation-induced DNA double-strand breaks. Stem Cells Dev 7:1471–1483
97. Fan J, Robert C, Jang YY, Liu H, Sharkis S, Baylin SB, Rassool FV (2011) Human induced pluripotent cells resemble embryonic stem cells demonstrating enhanced levels of DNA repair and efficacy of nonhomologous end-joining. Mutat Res 713:8–17
98. Adams BR, Golding SE, Rao RR, Valerie K (2010) Dynamic dependence on ATR and ATM for double-strand break repair in human embryonic stem cells and neural descendants. PLoS One 5:e10001
99. Fung H, Weinstock DM (2011) Repair at single targeted DNA double-strand breaks in pluripotent and differentiated human cells. PLoS One 6:e20514
100. Adams BR, Hawkins AJ, Povirk LF, Valerie K (2010) ATM-independent, high-fidelity nonhomologous end joining predominates in human embryonic stem cells. Aging (Albany NY) 2:582–596
101. Finnie NJ, Gottlieb TM, Blunt T, Jeggo PA, Jackson SP (1995) DNA-dependent protein kinase activity is absent in xrs-6 cells: implications for site-specific recombination and DNA double-strand break repair. Proc Natl Acad Sci USA 92:320–324
102. Anderson CW, Lees-Miller SP (1992) The nuclear serine/threonine protein kinase DNA-PK. Crit Rev Eukaryot Gene Expr 2:283–314
103. Li G, Nelsen C, Hendrickson EA (2002) Ku86 is essential in human somatic cells. Proc Natl Acad Sci USA 99:832–837
104. Van Sloun PP, Jansen JG, Weeda G, Mullenders LH, van Zeeland AA, Lohman PH, Vrieling H (1999) The role of nucleotide excision repair in protecting embryonic stem cells from genotoxic effects of UV-induced DNA damage. Nucl Acids Res 27:3276–3282
105. de Waard H, Sonneveld E, de Wit J, Esveldt-van Lange R, Hoeijmakers JH, Vrieling H, van Der Horst GT (2008) Cell-type-specific consequences of nucleotide excision repair deficiencies: embryonic stem cells versus fibroblasts. DNA Repair (Amst) 7:1659–1669
106. Tichy ED, Liang L, Deng L, Tischfield J, Schwemberger S, Babcock G, Stambrook PJ (2011) Mismatch and base excision repair proficiency in murine embryonic stem cells. DNA Repair (Amst) 10:445–451
107. Roos WP, Christmann M, Fraser ST, Kaina B (2007) Mouse embryonic stem cells are hypersensitive to apoptosis triggered by the DNA damage O(6)-methylguanine due to high E2F1 regulated mismatch repair. Cell Death Differ 14:1422–1432
108. Kim K, Doi A, Wen B, Ng K, Zhao R, Cahan P, Kim J, Aryee MJ, Ji H, Ehrlich LI, Yabuuchi A, Takeuchi A, Cunniff KC, Hongguang H, McKinney-Freeman S, Naveiras O, Yoon TJ, Irizarry RA, Jung N, Seita J, Hanna J, Murakami P, Jaenisch R, Weissleder R, Orkin SH, Weissman IL, Feinberg AP, Daley GQ (2010) Epigenetic memory in induced pluripotent stem cells. Nature 467:285–290
109. Hong H, Takahashi K, Ichisaka T, Aoi T, Kanagawa O, Nakagawa M, Okita K, Yamanaka S (2009) Suppression of induced pluripotent stem cell generation by the p53-p21 pathway. Nature 460:1132–1135
110. Kawamura T, Suzuki J, Wang YV, Menendez S, Morera LB, Raya A, Wahl GM, Belmonte JC (2009) Linking the p53 tumour suppressor pathway to somatic cell reprogramming. Nature 460:1140–1144
111. Li H, Collado M, Villasante A, Strati K, Ortega S, Canamero M, Blasco MA, Serrano M (2009) The Ink4/Arf locus is a barrier for iPS cell reprogramming. Nature 460:1136–1139
112. Marion RM, Strati K, Li H, Murga M, Blanco R, Ortega S, Fernandez-Capetillo O, Serrano M, Blasco MA (2009) A p53-mediated DNA damage response limits reprogramming to ensure iPS cell genomic integrity. Nature 460:1149–1153
113. Utikal J, Polo JM, Stadtfeld M, Maherali N, Kulalert W, Walsh RM, Khalil A, Rheinwald JG, Hochedlinger K (2009) Immortalization eliminates a roadblock during cellular reprogramming into iPS cells. Nature 460:1145–1148
114. Chao C, Wu Z, Mazur SJ, Borges H, Rossi M, Lin T, Wang JY, Anderson CW, Appella E, Xu Y (2006) Acetylation of mouse p53 at lysine 317 negatively regulates p53 apoptotic activities after DNA damage. Mol Cell Biol 26:6859–6869

115. Di Giovanni S, Knights CD, Rao M, Yakovlev A, Beers J, Catania J, Avantaggiati ML, Faden AI (2006) The tumor suppressor protein p53 is required for neurite outgrowth and axon regeneration. Embo J 25:4084–4096
116. Gil-Perotin S, Haines JD, Kaur J, Marin-Husstege M, Spinetta MJ, Kim KH, Duran-Moreno M, Schallert T, Zindy F, Roussel MF, Garcia-Verdugo JM, Casaccia P (2011) Roles of p53 and p27(Kip1) in the regulation of neurogenesis in the murine adult subventricular zone. Eur J Neurosci 34:1040–1052
117. Gil-Perotin S, Marin-Husstege M, Li J, Soriano-Navarro M, Zindy F, Roussel MF, Garcia-Verdugo JM, Casaccia-Bonnefil P (2006) Loss of p53 induces changes in the behavior of subventricular zone cells: implication for the genesis of glial tumors. J Neurosci 26:1107–1116
118. Zheng H, Ying H, Yan H, Kimmelman AC, Hiller DJ, Chen AJ, Perry SR, Tonon G, Chu GC, Ding Z, Stommel JM, Dunn KL, Wiedemeyer R, You MJ, Brennan C, Wang YA, Ligon KL, Wong WH, Chin L, DePinho RA (2008) p53 and Pten control neural and glioma stem/progenitor cell renewal and differentiation. Nature 455:1129–1133
119. Sherman MH, Kuraishy AI, Deshpande C, Hong JS, Cacalano NA, Gatti RA, Manis JP, Damore MA, Pellegrini M, Teitell MA (2010) AID-induced genotoxic stress promotes B cell differentiation in the germinal center via ATM and LKB1 signaling. Mol Cell 39:873–885
120. Liu Y, Elf SE, Miyata Y, Sashida G, Huang G, Di Giandomenico S, Lee JM, Deblasio A, Menendez S, Antipin J, Reva B, Koff A, Nimer SD (2009) p53 regulates hematopoietic stem cell quiescence. Cell Stem Cell 4:37–48
121. Liu Y, Hoya-Arias R, Nimer SD (2009) The role of p53 in limiting somatic cell reprogramming. Cell Res 19:1227–1228
122. Deng W, Xu Y (2009) Genome integrity: linking pluripotency and tumorgenicity. Trends Genet 25:425–427
123. Hussein SM, Batada NN, Vuoristo S, Ching RW, Autio R, Narva E, Ng S, Sourour M, Hamalainen R, Olsson C, Lundin K, Mikkola M, Trokovic R, Peitz M, Brustle O, Bazett-Jones DP, Alitalo K, Lahesmaa R, Nagy A, Otonkoski T (2011) Copy number variation and selection during reprogramming to pluripotency. Nature 471:58–62
124. Laurent LC, Ulitsky I, Slavin I, Tran H, Schork A, Morey R, Lynch C, Harness JV, Lee S, Barrero MJ, Ku S, Martynova M, Semechkin R, Galat V, Gottesfeld J, Izpisua Belmonte JC, Murry C, Keirstead HS, Park HS, Schmidt U, Laslett AL, Muller FJ, Nievergelt CM, Shamir R, Loring JF (2011) Dynamic changes in the copy number of pluripotency and cell proliferation genes in human ESCs and iPSCs during reprogramming and time in culture. Cell Stem Cell 8:106–118
125. Mayshar Y, Ben-David U, Lavon N, Biancotti JC, Yakir B, Clark AT, Plath K, Lowry WE, Benvenisty N (2010) Identification and classification of chromosomal aberrations in human induced pluripotent stem cells. Cell Stem Cell 7:521–531
126. Gore A, Li Z, Fung HL, Young JE, Agarwal S, Antosiewicz-Bourget J, Canto I, Giorgetti A, Israel MA, Kiskinis E, Lee JH, Loh YH, Manos PD, Montserrat N, Panopoulos AD, Ruiz S, Wilbert ML, Yu J, Kirkness EF, Izpisua Belmonte JC, Rossi DJ, Thomson JA, Eggan K, Daley GQ, Goldstein LS, Zhang K (2011) Somatic coding mutations in human induced pluripotent stem cells. Nature 471:63–67
127. Pera MF (2011) Stem cells: the dark side of induced pluripotency. Nature 471:46–47
128. Quinlan AR, Boland MJ, Leibowitz ML, Shumilina S, Pehrson SM, Baldwin KK, Hall IM (2011) Genome sequencing of mouse induced pluripotent stem cells reveals retroelement stability and infrequent DNA rearrangement during reprogramming. Cell Stem Cell 9:366–373
129. Fuchs E (2009) The tortoise and the hair: slow-cycling cells in the stem cell race. Cell 137:811–819
130. Facucho-Oliveira JM, St John JC (2009) The relationship between pluripotency and mitochondrial DNA proliferation during early embryo development and embryonic stem cell differentiation. Stem Cell Rev 5:140–158
131. Naka K, Muraguchi T, Hoshii T, Hirao A (2008) Regulation of reactive oxygen species and genomic stability in hematopoietic stem cells. Antioxid Redox Signal 10:1883–1894

132. Rehman J (2010) Empowering self-renewal and differentiation: the role of mitochondria in stem cells. J Mol Med (Berl) 88:981–986
133. Mohrin M, Bourke E, Alexander D, Warr MR, Barry-Holson K, Le Beau MM, Morrison CG, Passegue E (2010) Hematopoietic stem cell quiescence promotes error-prone DNA repair and mutagenesis. Cell Stem Cell 7:174–185
134. Nijnik A, Woodbine L, Marchetti C, Dawson S, Lambe T, Liu C, Rodrigues NP, Crockford TL, Cabuy E, Vindigni A, Enver T, Bell JI, Slijepcevic P, Goodnow CC, Jeggo PA, Cornall RJ (2007) DNA repair is limiting for haematopoietic stem cells during ageing. Nature 447:686–690
135. Rossi DJ, Bryder D, Seita J, Nussenzweig A, Hoeijmakers J, Weissman IL (2007) Deficiencies in DNA damage repair limit the function of haematopoietic stem cells with age. Nature 447:725–729
136. Mandal PK, Blanpain C, Rossi DJ (2011) DNA damage response in adult stem cells: pathways and consequences. Nat Rev Mol Cell Biol 12:198–202
137. Botchkarev VA, Komarova EA, Siebenhaar F, Botchkareva NV, Komarov PG, Maurer M, Gilchrest BA, Gudkov AV (2000) p53 is essential for chemotherapy-induced hair loss. Cancer Res 60:5002–5006
138. Song S, Lambert PF (1999) Different responses of epidermal and hair follicular cells to radiation correlate with distinct patterns of p53 and p21 induction. Am J Pathol 155:1121–1127
139. Diehn M, Cho RW, Lobo NA, Kalisky T, Dorie MJ, Kulp AN, Qian D, Lam JS, Ailles LE, Wong M, Joshua B, Kaplan MJ, Wapnir I, Dirbas FM, Somlo G, Garberoglio C, Paz B, Shen J, Lau SK, Quake SR, Brown JM, Weissman IL, Clarke MF (2009) Association of reactive oxygen species levels and radioresistance in cancer stem cells. Nature 458:780–783
140. Moore N, Lyle S (2011) Quiescent, slow-cycling stem cell populations in cancer: a review of the evidence and discussion of significance. J Oncol 2011
141. Niedernhofer LJ (2008) DNA repair is crucial for maintaining hematopoietic stem cell function. DNA Repair (Amst) 7:523–529
142. Milyavsky M, Gan OI, Trottier M, Komosa M, Tabach O, Notta F, Lechman E, Hermans KG, Eppert K, Konovalova Z, Ornatsky O, Domany E, Meyn MS, Dick JE (2010) A distinctive DNA damage response in human hematopoietic stem cells reveals an apoptosis-independent role for p53 in self-renewal. Cell Stem Cell 7:186–197
143. Sotiropoulou PA, Candi A, Mascre G, De Clercq S, Youssef KK, Lapouge G, Dahl E, Semeraro C, Denecker G, Marine JC, Blanpain C (2010) Bcl-2 and accelerated DNA repair mediates resistance of hair follicle bulge stem cells to DNA-damage-induced cell death. Nat Cell Biol 12:572–582
144. Merritt AJ, Potten CS, Kemp CJ, Hickman JA, Balmain A, Lane DP, Hall PA (1994) The role of p53 in spontaneous and radiation-induced apoptosis in the gastrointestinal tract of normal and p53-deficient mice. Cancer Res 54:614–617
145. Merritt AJ, Potten CS, Watson AJ, Loh DY, Nakayama K, Hickman JA (1995) Differential expression of bcl-2 in intestinal epithelia. Correlation with attenuation of apoptosis in colonic crypts and the incidence of colonic neoplasia. J Cell Sci 108(Pt 6):2261–2271
146. Miyamoto K, Araki KY, Naka K, Arai F, Takubo K, Yamazaki S, Matsuoka S, Miyamoto T, Ito K, Ohmura M, Chen C, Hosokawa K, Nakauchi H, Nakayama K, Nakayama KI, Harada M, Motoyama N, Suda T, Hirao A (2007) Foxo3a is essential for maintenance of the hematopoietic stem cell pool. Cell Stem Cell 1:101–112
147. Tothova Z, Kollipara R, Huntly BJ, Lee BH, Castrillon DH, Cullen DE, McDowell EP, Lazo-Kallanian S, Williams IR, Sears C, Armstrong SA, Passegue E, DePinho RA, Gilliland DG (2007) FoxOs are critical mediators of hematopoietic stem cell resistance to physiologic oxidative stress. Cell 128:325–339
148. Yalcin S, Zhang X, Luciano JP, Mungamuri SK, Marinkovic D, Vercherat C, Sarkar A, Grisotto M, Taneja R, Ghaffari S (2008) Foxo3 is essential for the regulation of ataxia telangiectasia mutated and oxidative stress-mediated homeostasis of hematopoietic stem cells. J Biol Chem 283:25692–25705

149. Ito K, Hirao A, Arai F, Matsuoka S, Takubo K, Hamaguchi I, Nomiyama K, Hosokawa K, Sakurada K, Nakagata N, Ikeda Y, Mak TW, Suda T (2004) Regulation of oxidative stress by ATM is required for self-renewal of haematopoietic stem cells. Nature 431:997–1002
150. Ito K, Hirao A, Arai F, Takubo K, Matsuoka S, Miyamoto K, Ohmura M, Naka K, Hosokawa K, Ikeda Y, Suda T (2006) Reactive oxygen species act through p38 MAPK to limit the lifespan of hematopoietic stem cells. Nat Med 12:446–451
151. Grinstein E, Mahotka C (2009) Stem cell divisions controlled by the proto-oncogene BMI-1. J Stem Cells 4:141–146
152. Liu J, Cao L, Chen J, Song S, Lee IH, Quijano C, Liu H, Keyvanfar K, Chen H, Cao LY, Ahn BH, Kumar NG, Rovira II, Xu XL, van Lohuizen M, Motoyama N, Deng CX, Finkel T (2009) Bmi1 regulates mitochondrial function and the DNA damage response pathway. Nature 459:387–392
153. Facchino S, Abdouh M, Chatoo W, Bernier G (2010) BMI1 confers radioresistance to normal and cancerous neural stem cells through recruitment of the DNA damage response machinery. J Neurosci 30:10096–10111
154. Mohyeldin A, Garzon-Muvdi T, Quinones-Hinojosa A (2010) Oxygen in stem cell biology: a critical component of the stem cell niche. Cell Stem Cell 7:150–161

Chapter 5
DNA Repair Mechanisms in Glioblastoma Cancer Stem Cells

Monica Venere, Jeremy N. Rich and Shideng Bao

Abstract Glioblastomas remain the most common and deadly adult brain tumor despite numerous advances made in the understanding of tumor biology. One such advance is the recent appreciation for a cellular hierarchy within the tumor bulk with only a subpopulation of cells, termed cancer stem cells, able to reinitiate tumor growth in transplantation assays. With the identification of these cells comes a further complexity in our consideration of how the heterogeneous cell populations within the tumor respond to therapeutic intervention. Cancer stem cells within glioblastomas, or glioma stem cells (GSCs), have been reported to have a chemo- and radioresistance phenotype as compared to the non-stem cell population. This is critical for patient care as radiotherapy and chemotherapy with the DNA alkylating agent, temozolomide, are the current standards of care. Importantly, both of these treatments rely on the cellular response to DNA damage to elicit their therapeutic benefit. For GSCs, the field is just beginning to appreciate how these cells respond on a molecular level to DNA damage. Nonetheless, advances have been made that highlight novel modes of potential therapeutic intervention and underscore the requirement for further studies aimed at elucidation of this key cellular pathway in GSC biology. This chapter summarizes our current understanding of the DNA damage response in GSCs.

5.1 Introduction

Gliomas are the most common primary brain tumors that are characterized by shared histological features with non-neuronal cell types within the brain. The main types of gliomas include: ependymomas, astrocytomas, oligodendrogliomas and mixed gliomas (multiple types of glial tumor cells represented). Further pathological classification of gliomas is most commonly reported using the World Health Organization (WHO) grading system, which categorizes gliomas into Grades I–IV, within a defined histology. Grade I glioma represents the least advanced disease and the best prognosis whereas Grade IV astrocytoma, also known as Glioblastoma Multiforme (GBM), is the most malignant form of the disease and is associated with an extremely

S. Bao (✉) · M. Venere · J. N. Rich
Department of Stem Cell Biology and Regenerative Medicine, Lerner Research Institute, Cleveland Clinic, Cleveland, OH, USA
e-mail: baos@ccf.org

L. A. Mathews et al. (eds.), *DNA Repair of Cancer Stem Cells*,
DOI 10.1007/978-94-007-4590-2_5, © Springer Science+Business Media Dordrecht 2013

poor prognosis. The median survival time of patients with GBM is approximately 12–15 months making GBM the most lethal as well as the most common adult brain tumor [1]. GBM can also be classified as primary or secondary based on the presence of a preceding low-grade lesion. Primary GBMs represent the majority of cases ($\sim$60 %) and usually present in older individuals (>50 years) following a very short clinical history, usually less than a few months. Secondary GBMs typically progress from a WHO Grade II or III glioma taking anywhere from 1 year to 10 years to present as a GBM and usually develop in younger patients (<45 years).

Numerous advances have been made in recent years toward understanding tumor biology such as the requirement of angiogenesis for tumor growth, immune evasion by tumor cells, as well as common mutations within tumor types. For GBM, the most common genetic alterations include: loss of heterozygosity (LOH) on chromosome arm 10q, mutations in the *TP53* tumor suppressor gene, amplification and activating rearrangements of the Epidermal Growth Factor Receptor (*EGFR*) gene, amplification or overexpression of the Human Double Minute (*HDM2*) gene, overexpression of Platelet-Derived Growth Factor-alpha (*PDGFA*) as well as overexpression and activating rearrangements of the Platelet-Derived Growth Factor Receptor-alpha (*PDGFRA*) gene, and mutations in the Phosphatase and TENsin homolog (*PTEN*) tumor suppressor gene [2]. A more in-depth analysis of common genetic alteration in GBM, performed through The Cancer Genome Atlas (TCGA), has stratified the disease into four subtypes: (1) classical (amplifications and mutations in the *EGFR* gene), (2) proneural (mutations in the *TP53* gene, mutations/amplifications in the *PDGFRA* gene, and point mutations in the *IDH1* gene), (3) mesenchymal (deletions in the *NF1* tumor suppressor gene), and (4) neural (expression changes in the neural-related *GABRA1* and *SLC12A5* genes) [3]. These subtypes were linked to patient response to intensive therapy with classical and mesenchymal subtypes having the greatest benefit, limited efficacy in the neural subtype, and no survival benefit in the proneural subytpe.

Understanding alterations in the genetic landscape of the tumor bulk and potential contributions to the therapeutic response are extremely important, but perhaps greater functional relevance for tumor recurrence following intervention can be found in the cellular hierarchy identified for GBM. The hierarchal model for the architecture of tumors posits that not all cells within the tumor have the same inherent tumorigenic potential when challenged to reform a phenotypic copy of the primary human lesion in immunocompromised mouse models or when functionally tested for self-renewal and proliferation. This model counters the stochastic clonal expansion model, whereby it is thought that any cell within the tumor has the same potential to reform that tumor. The cells at the apex of the GBM tumor hierarchy have been termed glioma stem cells (GSCs), or tumor propagating/initiating cells, and have been validated by numerous groups [4–10]. We will first describe how GSCs are isolated and functionally verified. We will then summarize the evidence that these cells are resistant to radiation and chemotherapy. Finally, we will evaluate what is known about the DNA Damage Response (DDR) in this subpopulation and how alterations in DDR pathways contribute to the resistance phenotype.

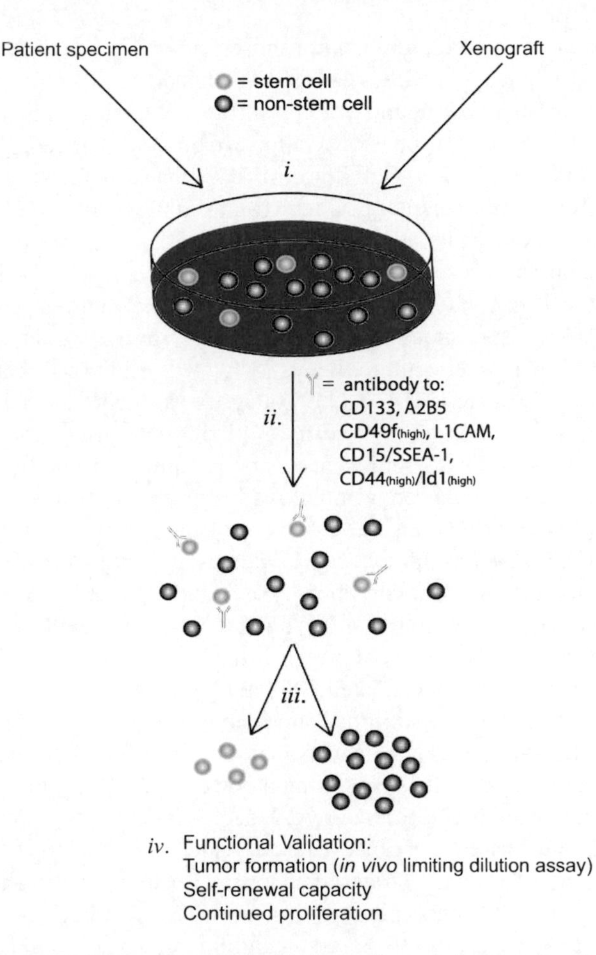

Fig. 5.1 Enrichment of Glioma Stem Cells. (i) Tumor tissue is isolated directly from a patient specimen or a xenograft tumor and dissociated into a single cell suspension. it (ii) The cells are then incubated with a fluorescently labeled antibody specific for a validated stem cell marker [e.g.; CD133, A2B5, CD49f(high), L1CAM, CD15/SSEA-1, CD44(high)/Id1(high)]. it (iii) The labeled cells are then passed through a Fluorescently Activated Cell Sorter (FACS) and collected into marker positive and marker negative fractions. (iv) The marker positive cells are functionally validated as the putative stem cell population as compared to the marker negative cells

5.2 Enrichment of Glioma Stem Cells

To isolate GSCs from a tumor, cell surface receptors are used to prospectively enrich or deplete for GSCs from the bulk tumor cell population (Fig. 5.1). This approach begins with the dissociation of the primary human tumor specimen into a single cell suspension. Central to the hierarchal tumor model is that all enrichment markers must be identified and validated from a primary patient specimen. However, further studies are often accomplished through the use of xenograft systems whereby the human cells are propagated in an immunocompromised mouse model with the hierarchy remaining intact whereby the GSCs can be re-isolated from the xenograft tumor. A fluorescently tagged antibody against the cell surface receptor, or marker, of interest is then incubated with the cell suspension to allow for binding. The cells are then passed through a Fluorescent Activated Cell Sorter (FACS) that will allow for identification of the cells positive for the receptor of interest and simultaneous sorting of these cells into the marker positive and marker negative fractions.

The most commonly used surface marker to enrich for GSCs is CD133 (or Prominin-1), a cell membrane glycoprotein, which has been repeatedly validated from primary patient specimens, although some controversy has arisen from studies performed on extensively cultured cells [4, 9, 10]. Additional markers include: CD15/SSEA-1, A2B5, L1CAM, CD49f/integrin α6(high) and CD44(high)/Id1 (high) [11–15]. It is important to note that utilizing any of the above markers to enrich for GSCs using FACS represents marker expression during a snapshot in time, meaning that the tumor environment is a dynamic one and a cell may gain or lose marker expression during development of the disease, especially for those markers whose expression is influenced by changes in the tumor microenvironment. This potential fluctuation in immunophenotype is of particular importance as the list of validated markers for GSCs grows. It is unlikely that GSCs will express all markers simultaneously and it is therefore possible that a marker negative tumor cell may in fact be positive for another validated GSC marker, potentially altering functional validation experiments. It is also important to note that alternative methods to sorting based on cell surface markers have been employed to enrich for tumor initiating cells from gliomas. Specifically, cell sorting based on a side population that is able to exclude the DNA-staining fluorescent dye, Hoechst 33342, has identified a glioma cell population with shared functional characteristics to GSCs [16, 17]. However, a contradictory report has demonstrated that sorting on side population is not necessary or sufficient for GSC enrichment [18]. Alternatively, tumorigenic glioma cells have been isolated by sorting cells based on autofluorescence (excitation at 480 nm, emission at 520 nm), collecting those cells within the high forward scatter and low side scatter populations [19]. Finally, some groups use spheroid culture conditions, without any prior marker sorting, to grow putative stem cells. However, this approach is extremely limiting as it relies solely on *in vitro* selection conditions to produce a potentially heterogeneous population of cells which can skew functional validation. Undeniably, no one marker or technical approach is without caveats and it is only through functional examination of the putative GSCs versus the non-GSC population that the hierarchy can be appreciated.

5.3 Functional Validation of Glioma Stem Cells

Central to the functional validation of GSCs is that these cells, as compared to the non-GSCs, can initiate tumor formation with a resulting tumor that is histologically similar to the patient tumor. Therefore, the most important experiment required to validate putative GSCs is an *in vivo* limiting dilution assay. This assay involves injecting a decreasing number of GSCs or non-GSCs, done in parallel with the same numbers, into an orthotopic host location, which for GBM would be intracranially. This identifies the minimal number of cells required for tumor initiation. Although tumor formation from a single cell injection has yet to be validated for GBM, it has been demonstrated that just a few hundred GSCs can lead to tumor formation whereas the non-GSCs consistently lack this ability in most cases [4, 9].

Additionally, GSCs must be able to self-renew and continually proliferate as a means to maintain and propagate the tumor. These functional hallmarks are most commonly monitored under cell culture conditions. Namely, serial passage of GSCs grown under conditions that promote the cells to form spheroids (or tumorspheres) is used as a surrogate for self-renewal and to monitor continued proliferation. A single GSC can grow into a tumorsphere, be dissociated, with a single GSC from that primary sphere able to reform a secondary tumorsphere and so on without exhaustion of the self-renewal capability or termination of the proliferative capacity. Additional experimental paradigms supporting GSC isolation include quantitative PCR for stem cell marker expression (e.g.; Sox2, Olig2) and the ability to respond to culture conditions favoring multi-lineage differentiation into neurons, astrocytes and/or oligodendrocytes. Importantly, GSCs from these culture assays should consistently maintain their tumor initiating phenotype. Although culture undoubtedly confers selection as well as other potential caveats, the above assays are key to GSC validation.

5.4 Response of Glioma Stem Cells to Current Therapies Targeting the DNA Damage Response

Currently, the standard of care for patients presenting with GBM is surgical resection followed by concurrent radiotherapy and adjuvant chemotherapy with the DNA alkylating agent temozolomide (TMZ). It is important to note that nearly all patients are refractory to these treatments and will experience disease recurrence. The poor survival outcome despite clinical intervention suggests that a population of cells within the tumor is capable of evading therapy-induced cell death. More specifically, these cells must harbor a phenotype that is resistant to cell death driven by DNA damage as both radiation and TMZ impinge directly on the DDR. Recent data indicates that GSCs demonstrate these evasion properties.

TMZ is a DNA damaging agent that works mainly by alkylating the O^6 position of guanine which, if left unrepaired, can lead to a DNA base mismatch and ultimately cell death. Repair of the alkylated guanine occurs by the O-6-methylguanine-DNA methyltransferase (MGMT) enzyme that will irreversibly transfer the alkyl group to a cysteine residue within the protein. The response of patients to TMZ has been linked to the methylation status of the MGMT gene, with hypermethylation correlating to partial or complete response in primary presentation of the disease [20, 21]. There have been conflicting reports in the literature regarding the response of GSCs to TMZ [16, 22–27]. Part of the controversy stems from the inability to directly compare putative stem cells from cell lines to primary patient derived stem cell cultures as well as overall variation in culture conditions and experimental design. Nonetheless, when higher expression of MGMT was described for GSCs there was a correlating resistance to TMZ whereas those GSCs without MGMT could be targeted by TMZ treatment. These results would indicate that GSCs lacking MGMT are sensitive to the resulting unsuccessful attempts at mismatch repair which ultimately lead to the formation of double strand breaks. However, more in depth studies are required to fully understand the response of GSCs lacking MGMT to TMZ on a molecular level.

As mentioned, radiation is given concurrently with TMZ during therapeutic intervention. It is usually administered in fractionated doses of 1.8–2.0 Gy (Gray; unit of radiation) daily (Monday to Friday) for five to seven weeks for a total administered dose of about 60 Gy [28]. External-beam radiation therapy (EBRT), which is used to focus the irradiation to the area of the tumor, induces double strand breaks and activation of the DDR. The goal of this treatment is to further eradicate any remaining post-resection tumor cells. Unfortunately, recurrence is nearly universal and with recent studies revealing their radioresistance phenotype, GSCs have been implicated in initiating tumor regrowth. Following ionizing radiation, the number of CD133 positive GSCs increased in both *in vitro* and *in vivo* model systems [4]. The biological significance of this enrichment was confirmed through a resulting decrease in latency and increase in tumor size that correlated with an increasing percentage of intracranially transplanted CD133 positive cells. Importantly, irradiated CD133 positive GSCs were able to generate tumors to nearly the same extent as non-irradiated CD133 positive GSCs, confirming their tumorigenic potential post-treatment. This expansion of CD133 positive cells following irradiation was confirmed in patients that underwent EBRT plus Gamma Knife surgery (GKS; non-invasive radiosurgery) [29]. In this study, 32 patients presenting with primary GBM underwent surgical resection along with GKS in combination with EBRT. Twelve patients that experienced recurrence again underwent surgical removal of the tumor. The authors were able to compare the percentage of CD133 positive cells in the primary and recurrent specimens from the same patients. The primary specimens had little to no CD133 positive cells whereas the mean percentage of CD133 positive tumor cells following treatment was nearly 17 %. Mechanistically, it appears that the CD133 positive GSCs have a higher basal activation of components of the DDR as well as more pronounced DDR activation following irradiation that contribute to greater repair kinetics over the non-GSCs [4]. Additionally, following irradiation, the GSCs were shown to undergo apoptosis less frequently than non-GSCs [4]. Together, these studies highlight a radioresistance phenotype for the GSCs and a potential direct contribution to tumor recurrence.

5.5 Role of the Microenvironment on the Response of Glioma Stem Cells to DNA Damage

It is now appreciated that GSCs reside within certain microenvironments, or niches, within the tumor. Namely, they are enriched around areas of hypoxia as well as the tumor vessels and these niches act to maintain the GSC phenotype [30–32].

Importantly, both of these tumor microenvironments (perivascular and hypoxic niches) impact the response of cells to radiation therapy. In a mouse model of medulloblastoma, it was shown that cancer stem cells in this system residing in the perivascular niche demonstrate increased activation of the Akt pathway as well as undergo a p53 dependent cell cycle arrest following irradiation [33]. These cells not only survive but also reenter the cell cycle whereas proliferating cells in the tumor bed undergo apoptosis [33]. Interestingly, the authors demonstrated that this radioresistant phenotype of the perivascular cancer stem cells is impaired through inhibition of the Akt pathway. Notch signaling within the perivascular niche has also been reported to be a key regulator of the GSC phenotype and inhibition of this pathway has demonstrated therapeutic potential by increasing radiosensitivity of GSCs [34–36]. These initial findings highlight a complex relationship between tumor vasculature and GSCs, underscoring the need for additional studies at the molecular level to further elucidate the impact of the perivascular niche on the DDR.

Hypoxic areas within tumors have long been known to modify the response of cells to radiation conferring 2–3 times more resistance [37]. Hypoxia has been shown to regulate the phenotype of GSCs with the hypoxia inducible factors, HIF2α, HIF1α and HIF1β, being central regulators of GSC function [31, 32, 38–40]. Importantly, survival of mice bearing intracranial tumors was extended when HIF2α and HIF1α were depleted in GSCs, suggesting that modulation of hypoxia induced regulation of these cells directly impacts their tumorigenicity [32]. However, the contribution of GSC residency within the hypoxic niche to their radioresistant phenotype has not been investigated, although a direct link is likely to exist. As the CSC field moves forward in attempting to understand the DDR in GSCs it will be important to take into consideration the tumor microenvironment and how best to recapitulate these conditions in studies as differential responses of GSCs to irradiation *in vitro* and *in vivo* [41].

5.6 DNA Repair and Checkpoint Activation in Glioma Stem Cells

The main DNA damage repair pathways (covered in detail in a Chap. 2) can be divided into two groups, (1) those that handle single-strand damage [base excision repair (BER), nucleotide excision repair (NER), and mismatch repair (MMR)] and, (2) those that deal with double-strand breaks [non-homologous end joining (NHEJ) and homologous recombination (HR)]. Exploration of the role of these pathways in GSCs is still in its infancy. Moreover, the field is lacking an appreciation of the *in vivo* regulation of the DDR on a molecular level. This is partially confounded by a lack of appreciation for the *in vivo* cell cycle regulation of GSCs. This is critical since the cell cycle has a major role in not only influencing the overall sensitivity to radiotherapy, but DNA damage is handled through differential repair mechanisms depending on cell cycle phase [42]. It has been reported that more quiescent cells within a GBM tumor have CSC phenotypes but this concept warrants further

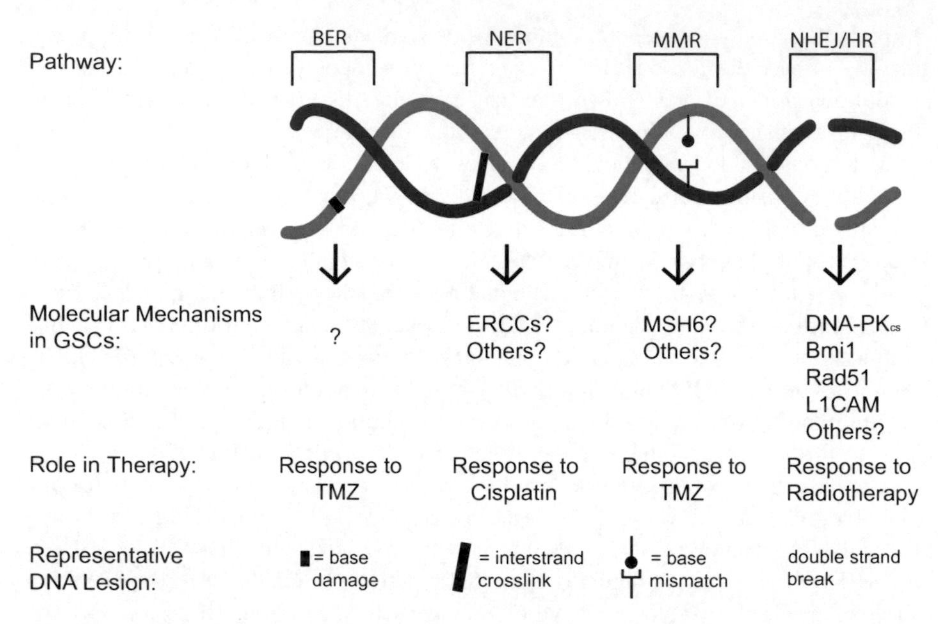

Fig. 5.2 DNA Repair Pathways in Glioma Stem Cells. The four major DNA repair pathways [Base Excision Repair (BER), Nucleotide Excision Repair (NER), Mismatch Repair (MMR) and double strand break repair, via Nonhomologous End Joining (NHEJ) or homologous recombination (HR)] have been minimally studied in glioma stem cells. Represented are the known or potential molecular mechanisms for each repair pathway as well as the main therapeutic intervention that each pathway would respond to during treatment

exploration [43]. Importantly, when GSCs are studied *in vitro*, they are maintained in growth promoting media [i.e.; supplemented with basic Fibroblast Growth Factor (bFGF) and Epidermal Growth Factor (EGF)], which would alter the phenotype of a potential quiescent GSC population. Nonetheless, cell intrinsic regulation of the DDR is likely to be maintained within CSCs and recent studies have revealed some important findings (Fig. 5.2).

5.6.1 Base Excision Repair and Nucleotide Excision Repair

The CSC field has yet to explore the role of BER and NER within GSCs. In general, NER in glioma is relatively unknown. A few studies have evaluated the status of various NER players within the Excision Repair Cross Complementing group of proteins (ERCCs) with the main finding that differential levels of these genes could influence the chemosensitivity of glioma cells to platinum-based therapies such as cisplatin, which require the NER pathway for resolution [44, 45]. A few groups have evaluated the response of GSCs to cisplatin treatment, revealing dichotomous results. In some instances, GSCs were resistant to treatment whereas other studies showed that cisplatin treatment decreased GSC viability and neurosphere formation [22, 46]. Clearly, additional work is required to understand NER on a molecular level in GSCs. BER has been minimally explored in GSCs. The Frosina lab utilized an *in vitro* BER assay whereby a DNA template containing a lesion requiring the BER machinery was incubated with extracts from both GSC and non-GSC lines with no difference in repair capacity observed between these two populations [47]. Interestingly, the majority of DNA lesions induced by TMZ are *N*-methylated bases that are recognized by DNA glycosylases, the central players in BER. Therefore, although resistance to TMZ is attributed to MGMT status and subsequent response to the O^6-methylguanine DNA lesions, it is possible that GSCs could be sensitized to TMZ, independent of MGMT status, through inhibition of BER. This hypothesis has been explored in glioma cell lines through direct depletion of BER enzymes as well as with drug inhibition of BER with promising results [48–50]. Furthermore, the combination of TMZ with an inhibitor of the BER enzyme, Poly(ADP-Ribose) polymerase-1 (PARP-1), inhibited tumor growth and prolonged animal survival in a orthotopic rat glioma model [51]. It is necessary to expand these findings to the GSC population as potential resistance of GSCs to TMZ, and subsequent recurrence by surviving tumor-initiating cells, might be altered through manipulation of BER.

5.6.2 Mismatch Repair

As touched on previously, TMZ treatment will lead to methylation of the O^6 position on guanine, normally removed by the MGMT repair enzyme. However, if not repaired by MGMT (i.e.; in cells deficient for this protein), it is proposed that the methylated guanine will lead to futile rounds of MMR and ultimately double strand breaks. Interestingly, loss of the MMR protein, MSH6, has been associated with progression of GBM following TMZ treatment and may contribute to drug resistance and tumor recurrence [52, 53]. Still unexplored, however, is the selection for MSH6 mutations within GSCs and the resulting direct contribution of these cells to TMZ resistance and relapse through this mechanism. Moreover, a more in depth profiling of the MMR pathway in GSCs is required to strengthen the understanding of the role of this repair mechanism in GSC biology.

5.6.3 Homologous Recombination and Non-Homologous End Joining

The major DNA lesions resulting from ionizing radiation are double strand breaks, repaired by HR or NHEJ. It is unknown as to which pathway is more dominant, if at all, in GSCs. As previously mentioned, there is a lack of information regarding the proliferation status of GSCs *in vivo*. As HR requires a homologous template to occur, a more quiescent state would favor NHEJ. Although the field would undoubtedly benefit from rigorous studies evaluating the cell cycle status along with the response to radiotherapy of GSCs *in vivo*, there have nonetheless been advances made using culture systems toward our understanding of overall GSC response to ionizing radiation as well as the role of certain key proteins known to be involved in HR or NHEJ.

A central player in repair via NHEJ is the kinase, DNA-PK_{cs}, along with the accessory heterodimeric complex of Ku70 and Ku80. Recently, it has been reported that knockdown of DNA-PK_{cs} by RNA interference in GSCs will radiosensitize these cells [54]. Although the authors did not evaluate the impact of DNA-PK_{cs} knockdown on DNA repair by NHEJ, they did identify an increase in autophagic cell death following irradiation. Autophagy is a cellular process that can be activated by chemotherapy or irradiation and can lead to cell death or resistance [55–59]. Previous studies had reported that autophagy by ionizing radiation led to a radioresistant phenotype in CD133 positive GSCs [60]. However, the role of DNA-PK_{cs} was not evaluated in this study and the recent data linking DNA-PK_{cs} depletion with increased authophagy induced cell death suggests that alterations in the ability to repair DNA by NHEJ could be deleterious to GSCs. Additional elucidation of the importance of NHEJ in GSCs was highlighted by the recent report that the GSC regulatory protein, Bmi1, interacts with DNA-PK_{cs}, as well as other DNA repair proteins, following irradiation [61, 62]. Bmi1 is a polycomb group protein, involved in repressing gene expression through histone modifications and chromatin compaction [63, 64]. It has been shown to have a role in GSC self-renewal as inactivation of Bmi1 resulted in depletion of the CD133 positive population through differentiation and apoptosis [61]. Although not explored in the GSC population nor directly linked to a deficiency in NHEJ, depletion of Bmi1 radiosensitized GBM cells [62]. Although the above results are intriguing, further studies are required to elucidate the exact role of NHEJ in GSC biology.

As with NHEJ, the field is just beginning to appreciate the role of HR in GSCs. One of the key mediators of HR is Rad51, which controls strand invasion and recombination during DNA repair via HR. A recent report utilized freshly isolated, patient derived GBM cell lines to demonstrate a dependence on Rad51 for cell survival, even without the addition of radiation treatment [65]. The cell lines used in this study did contain a CD133 positive population but were not a pure, validated GSC line, so further studies are required to verify the dependence on Rad51 in GSCs as well as other central players in HR, such as BRCA1.

5.6.4 Checkpoint Activation in Glioma Stem Cells

Upstream of the repair process involving Rad51 is a signaling cascade involving the recognition of the double strand breaks by the MRE11-RAD50-NBS1 (MRN) complex which will signal to the ataxia telangiectasia mutated (ATM) kinase to allow for cell cycle checkpoint activation through the Chk1 and Chk2 kinases and/or apoptosis via the p53 tumor suppressor. As previously mentioned, CD133 positive GSCs have a survival advantage over the CD133 negative population following irradiation [4]. This finding was correlated to preferential activation of proteins in the DNA damage checkpoint response, including ATM, Chk1 and Chk2, as well as an increase in DNA repair capacity in the GSCs [4]. Importantly, compromised Chk1 and Chk2 function could increase the radiosensitivity of CD133 positive GSCs as well as CD133 containing tumor neurospheres [4, 66]. More recent studies have partially elucidated the molecular mechanisms responsible for the preferential checkpoint response in GSCs. Namely, the GSC regulatory protein L1CAM was shown to be central to DNA damage checkpoint activation and repair in GSCs [67]. L1CAM is a transmembrane protein reported to regulate cell adhesion, survival, growth, migration and invasion [68, 69]. In GSCs, it was demonstrated to support survival and tumor growth [13]. Depletion of L1CAM compromised the DNA repair capacity in GSCs as well as decreased their survival following DNA damage [67]. This phenotype was directly linked to a role for the intracellular domain of L1CAM in regulating Nbs1 expression through c-Myc [67]. These studies not only reveal important insights into the GSC biological response to DNA damage but also highlight novel mechanisms to potentially therapeutically target these cells.

5.7 Future Directions

Although the field has begun to make headway in its understanding of the DNA damage response in GSCs, there is still so much to learn. Moving forward, a better understanding of the *in vivo* response of GSCs to current treatments will be instrumental in designing more effective therapies aimed at eradicating these resistant cells. This will be aided by interrogation on the molecular level of the numerous players integral to the main DNA repair pathways or checkpoint signaling. Furthermore, the DNA damage response as related to the microenvironment and location of GSCs must be defined. Another unexplored area is the impact of DNA damage on cellular differentiation of GSCs. It will be very interesting to determine how the DNA damage response activates molecular signaling to trigger cell differentiation in GSCs. This is a relatively unexplored outcome of DNA damage yet of major therapeutic value in targeting the CSC population [70]. Finally, as novel chemotherapeutics aimed at DNA repair mechanisms continue to develop, they must be evaluated in terms of the cellular hierarchy within glioma as to best serve patients by validating efficacy against the tumor initiating apex cell.

References

1. Stupp R, Hegi ME, Neyns B, Goldbrunner R, Schlegel U, Clement PM, Grabenbauer GG, Ochsenbein AF, Simon M, Dietrich PY et al (2010) Phase I/IIa study of cilengitide and temozolomide with concomitant radiotherapy followed by cilengitide and temozolomide maintenance therapy in patients with newly diagnosed glioblastoma. J Clin Oncol 28(16):2712–2718

2. Sathornsumetee S, Reardon DA, Desjardins A, Quinn JA, Vredenburgh JJ, Rich JN (2007) Molecularly targeted therapy for malignant glioma. Cancer 110(1):13–24
3. Verhaak RG, Hoadley KA, Purdom E, Wang V, Qi Y, Wilkerson MD, Miller CR, Ding L, Golub T, Mesirov JP et al (2010) Integrated genomic analysis identifies clinically relevant subtypes of glioblastoma characterized by abnormalities in PDGFRA, IDH1, EGFR, and NF1. Cancer Cell 17(1):98–110
4. Bao S, Wu Q, McLendon RE, Hao Y, Shi Q, Hjelmeland AB, Dewhirst MW, Bigner DD, Rich JN (2006) Glioma stem cells promote radioresistance by preferential activation of the DNA damage response. Nature 444(7120):756–760
5. Galli R, Binda E, Orfanelli U, Cipelletti B, Gritti A, De Vitis S, Fiocco R, Foroni C, Dimeco F, Vescovi A (2004) Isolation and characterization of tumorigenic, stem-like neural precursors from human glioblastoma. Cancer Res 64(19):7011–7021
6. Hemmati HD, Nakano I, Lazareff JA, Masterman-Smith M, Geschwind DH, Bronner-Fraser M, Kornblum HI (2003) Cancerous stem cells can arise from pediatric brain tumors. Proc Natl Acad Sci USA 100(25):15178–15183
7. Ignatova TN, Kukekov VG, Laywell ED, Suslov ON, Vrionis FD, Steindler DA (2002) Human cortical glial tumors contain neural stem-like cells expressing astroglial and neuronal markers in vitro. Glia 39(3):193–206
8. Singh SK, Clarke ID, Terasaki M, Bonn VE, Hawkins C, Squire J, Dirks PB (2003) Identification of a cancer stem cell in human brain tumors. Cancer Res 63(18):5821–5828
9. Singh SK, Hawkins C, Clarke ID, Squire JA, Bayani J, Hide T, Henkelman RM, Cusimano MD, Dirks PB (2004) Identification of human brain tumour initiating cells. Nature 432(7015):396–401
10. Wang R, Chadalavada K, Wilshire J, Kowalik U, Hovinga KE, Geber A, Fligelman B, Leversha M, Brennan C, Tabar V (2010) Glioblastoma stem-like cells give rise to tumour endothelium. Nature 468(7325):829–833
11. Son MJ, Woolard K, Nam DH, Lee J, Fine HA (2009) SSEA-1 is an enrichment marker for tumor-initiating cells in human glioblastoma. Cell Stem Cell 4(5):440–452
12. Ogden AT, Waziri AE, Lochhead RA, Fusco D, Lopez K, Ellis JA, Kang J, Assanah M, McKhann GM, Sisti MB et al (2008) Identification of $A2B5^{+}$ $CD133^{-}$ tumor-initiating cells in adult human gliomas. Neurosurgery 62(2):505–514; discussion 514–515
13. Bao S, Wu Q, Li Z, Sathornsumetee S, Wang H, McLendon RE, Hjelmeland AB, Rich JN (2008) Targeting cancer stem cells through L1CAM suppresses glioma growth. Cancer Res 68(15):6043–6048
14. Lathia JD, Gallagher J, Heddleston JM, Wang J, Eyler CE, Macswords J, Wu Q, Vasanji A, McLendon RE, Hjelmeland AB et al (2010) Integrin alpha 6 regulates glioblastoma stem cells. Cell Stem Cell 6(5):421–432
15. Anido J, Saez-Borderias A, Gonzalez-Junca A, Rodon L, Folch G, Carmona MA, Prieto-Sanchez RM, Barba I, Martinez-Saez E, Prudkin L et al (2010) TGF-beta Receptor Inhibitors Target the CD44(high)/Id1(high) Glioma-Initiating Cell Population in Human Glioblastoma. Cancer Cell 18(6):655–668
16. Bleau AM, Hambardzumyan D, Ozawa T, Fomchenko EI, Huse JT, Brennan CW, Holland EC (2009) PTEN/PI3K/Akt pathway regulates the side population phenotype and ABCG2 activity in glioma tumor stem-like cells. Cell Stem Cell 4(3):226–235
17. Patrawala L, Calhoun T, Schneider-Broussard R, Zhou J, Claypool K, Tang DG (2005) Side population is enriched in tumorigenic, stem-like cancer cells, whereas $ABCG2^{+}$ and $ABCG2^{-}$ cancer cells are similarly tumorigenic. Cancer Res 65(14):6207–6219
18. Broadley KW, Hunn MK, Farrand KJ, Price KM, Grasso C, Miller RJ, Hermans IF, McConnell MJ (2011) Side population is not necessary or sufficient for a cancer stem cell phenotype in glioblastoma multiforme. Stem Cells 29(3):452–461
19. Clement V, Marino D, Cudalbu C, Hamou MF, Mlynarik V, de Tribolet N, Dietrich PY, Gruetter R, Hegi ME, Radovanovic I (2010) Marker-independent identification of glioma-initiating cells. Nat Methods 7(3):224–228

20. Hegi ME, Diserens AC, Gorlia T, Hamou MF, de Tribolet N, Weller M, Kros JM, Hainfellner JA, Mason W, Mariani L et al (2005) MGMT gene silencing and benefit from temozolomide in glioblastoma. N Engl J Med 352(10):997–1003
21. Paz MF, Yaya-Tur R, Rojas-Marcos I, Reynes G, Pollan M, Aguirre-Cruz L, Garcia-Lopez JL, Piquer J, Safont MJ, Balana C et al (2004) CpG island hypermethylation of the DNA repair enzyme methyltransferase predicts response to temozolomide in primary gliomas. Clin Cancer Res 10(15):4933–4938
22. Eramo A, Ricci-Vitiani L, Zeuner A, Pallini R, Lotti F, Sette G, Pilozzi E, Larocca LM, Peschle C, De Maria R (2006) Chemotherapy resistance of glioblastoma stem cells. Cell Death Differ 13(7):1238–1241
23. Pistollato F, Abbadi S, Rampazzo E, Persano L, Della Puppa A, Frasson C, Sarto E, Scienza R, D'Avella D, Basso G (2010) Intratumoral hypoxic gradient drives stem cells distribution and MGMT expression in glioblastoma. Stem Cells 28(5):851–862
24. Beier D, Rohrl S, Pillai DR, Schwarz S, Kunz-Schughart LA, Leukel P, Proescholdt M, Brawanski A, Bogdahn U, Trampe-Kieslich A et al (2008) Temozolomide preferentially depletes cancer stem cells in glioblastoma. Cancer Res 68(14):5706–5715
25. Mihaliak AM, Gilbert CA, Li L, Daou MC, Moser RP, Reeves A, Cochran BH, Ross AH (2010) Clinically relevant doses of chemotherapy agents reversibly block formation of glioblastoma neurospheres. Cancer Lett 296(2):168–177
26. Gilbert CA, Daou MC, Moser RP, Ross AH (2010) Gamma-secretase inhibitors enhance temozolomide treatment of human gliomas by inhibiting neurosphere repopulation and xenograft recurrence. Cancer Res 70(17):6870–6879
27. Clement V, Sanchez P, de Tribolet N, Radovanovic I, Ruiz i Altaba A (2007) HEDGEHOG-GLI1 signaling regulates human glioma growth, cancer stem cell self-renewal, and tumorigenicity. Curr Biol 17(2):165–172
28. Salacz ME, Watson KR, Schomas DA (2011) Glioblastoma: Part I. Current state of affairs. Mo Med 108(3):187–194
29. Tamura K, Aoyagi M, Wakimoto H, Ando N, Nariai T, Yamamoto M, Ohno K (2010) Accumulation of CD133-positive glioma cells after high-dose irradiation by Gamma Knife surgery plus external beam radiation. J Neurosurg 113(2):310–318
30. Calabrese C, Poppleton H, Kocak M, Hogg TL, Fuller C, Hamner B, Oh EY, Gaber MW, Finklestein D, Allen M et al (2007) A perivascular niche for brain tumor stem cells. Cancer Cell 11(1):69–82
31. McCord AM, Jamal M, Shankavaram UT, Lang FF, Camphausen K, Tofilon PJ (2009) Physiologic oxygen concentration enhances the stem-like properties of $CD133^+$ human glioblastoma cells in vitro. Mol Cancer Res 7(4):489–497
32. Li Z, Bao S, Wu Q, Wang H, Eyler C, Sathornsumetee S, Shi Q, Cao Y, Lathia J, McLendon RE et al (2009) Hypoxia-inducible factors regulate tumorigenic capacity of glioma stem cells. Cancer Cell 15(6):501–513
33. Hambardzumyan D, Becher OJ, Rosenblum MK, Pandolfi PP, Manova-Todorova K, Holland EC (2008) PI3K pathway regulates survival of cancer stem cells residing in the perivascular niche following radiation in medulloblastoma in vivo. Genes Dev 22(4):436–448
34. Charles N, Ozawa T, Squatrito M, Bleau AM, Brennan CW, Hambardzumyan D, Holland EC (2010) Perivascular nitric oxide activates notch signaling and promotes stem-like character in PDGF-induced glioma cells. Cell Stem Cell 6(2):141–152
35. Wang J, Wakeman TP, Lathia JD, Hjelmeland AB, Wang XF, White RR, Rich JN, Sullenger BA (2010) Notch promotes radioresistance of glioma stem cells. Stem Cells 28(1):17–28
36. Hovinga KE, Shimizu F, Wang R, Panagiotakos G, Van Der Heijden M, Moayedpardazi H, Correia AS, Soulet D, Major T, Menon J et al (2010) Inhibition of notch signaling in glioblastoma targets cancer stem cells via an endothelial cell intermediate. Stem Cells 28(6):1019–1029
37. Overgaard J (2007) Hypoxic radiosensitization: adored and ignored. J Clin Oncol 25(26):4066–4074
38. Soeda A, Park M, Lee D, Mintz A, Androutsellis-Theotokis A, McKay RD, Engh J, Iwama T, Kunisada T, Kassam AB et al (2009) Hypoxia promotes expansion of the CD133-positive glioma stem cells through activation of HIF-1alpha. Oncogene 28(45):3949–3959

39. Bar EE, Lin A, Mahairaki V, Matsui W, Eberhart CG (2010) Hypoxia increases the expression of stem-cell markers and promotes clonogenicity in glioblastoma neurospheres. Am J Pathol 177(3):1491–1502
40. Seidel S, Garvalov BK, Wirta V, von Stechow L, Schanzer A, Meletis K, Wolter M, Sommerlad D, Henze AT, Nister M et al (2010) A hypoxic niche regulates glioblastoma stem cells through hypoxia inducible factor 2 alpha. Brain 133(Pt 4):983–995
41. Jamal M, Rath BH, Williams ES, Camphausen K, Tofilon PJ (2010) Microenvironmental regulation of glioblastoma radioresponse. Clin Cancer Res 16(24):6049–6059
42. Pawlik TM, Keyomarsi K (2004) Role of cell cycle in mediating sensitivity to radiotherapy. Int J Radiat Oncol Biol Phys 59(4):928–942
43. Deleyrolle LP, Harding A, Cato K, Siebzehnrubl FA, Rahman M, Azari H, Olson S, Gabrielli B, Osborne G, Vescovi A et al (2011) Evidence for label-retaining tumour-initiating cells in human glioblastoma. Brain 134(Pt 5):1331–1343
44. Chen H, Shao C, Shi H, Mu Y, Sai K, Chen Z (2007) Single nucleotide polymorphisms and expression of ERCC1 and ERCC2 vis-a-vis chemotherapy drug cytotoxicity in human glioma. J Neurooncol 82(3):257–262
45. Chen HY, Shao CJ, Chen FR, Kwan AL, Chen ZP (2010) Role of ERCC1 promoter hypermethylation in drug resistance to cisplatin in human gliomas. Int J Cancer 126(8):1944–1954
46. Ding L, Yuan C, Wei F, Wang G, Zhang J, Bellail AC, Zhang Z, Olson JJ, Hao C (2011) Cisplatin restores TRAIL apoptotic pathway in glioblastoma-derived stem cells through up-regulation of DR5 and down-regulation of c-FLIP. Cancer Invest 29(8):511–520
47. Ropolo M, Daga A, Griffero F, Foresta M, Casartelli G, Zunino A, Poggi A, Cappelli E, Zona G, Spaziante R et al (2009) Comparative analysis of DNA repair in stem and nonstem glioma cell cultures. Mol Cancer Res 7(3):383–392
48. Trivedi RN, Almeida KH, Fornsaglio JL, Schamus S, Sobol RW (2005) The role of base excision repair in the sensitivity and resistance to temozolomide-mediated cell death. Cancer Res 65(14):6394–6400
49. Tang JB, Svilar D, Trivedi RN, Wang XH, Goellner EM, Moore B, Hamilton RL, Banze LA, Brown AR, Sobol RW (2011) N-methylpurine DNA glycosylase and DNA polymerase beta modulate BER inhibitor potentiation of glioma cells to temozolomide. Neuro Oncol 13(5):471–486
50. Tentori L, Portarena I, Torino F, Scerrati M, Navarra P, Graziani G (2002) Poly(ADP-ribose) polymerase inhibitor increases growth inhibition and reduces G(2)/M cell accumulation induced by temozolomide in malignant glioma cells. Glia 40(1):44–54
51. Donawho CK, Luo Y, Penning TD, Bauch JL, Bouska JJ, Bontcheva-Diaz VD, Cox BF, DeWeese TL, Dillehay LE, Ferguson DC et al (2007) ABT-888, an orally active poly(ADP-ribose) polymerase inhibitor that potentiates DNA-damaging agents in preclinical tumor models. Clin Cancer Res 13(9):2728–2737
52. Cahill DP, Levine KK, Betensky RA, Codd PJ, Romany CA, Reavie LB, Batchelor TT, Futreal PA, Stratton MR, Curry WT et al (2007) Loss of the mismatch repair protein MSH6 in human glioblastomas is associated with tumor progression during temozolomide treatment. Clin Cancer Res 13(7):2038–2045
53. Yip S, Miao J, Cahill DP, Iafrate AJ, Aldape K, Nutt CL, Louis DN (2009) MSH6 mutations arise in glioblastomas during temozolomide therapy and mediate temozolomide resistance. Clin Cancer Res 15(14):4622–4629
54. Zhuang W, Li B, Long L, Chen L, Huang Q, Liang ZQ (2011) Knockdown of the DNA-dependent protein kinase catalytic subunit radiosensitizes glioma-initiating cells by inducing autophagy. Brain Res 1371:7–15
55. Li J, Qin Z, Liang Z (2009) The prosurvival role of autophagy in Resveratrol-induced cytotoxicity in human U251 glioma cells. BMC Cancer 9:215
56. Paglin S, Hollister T, Delohery T, Hackett N, McMahill M, Sphicas E, Domingo D, Yahalom J (2001) A novel response of cancer cells to radiation involves autophagy and formation of acidic vesicles. Cancer Res 61(2):439–444

57. Yao KC, Komata T, Kondo Y, Kanzawa T, Kondo S, Germano IM (2003) Molecular response of human glioblastoma multiforme cells to ionizing radiation: cell cycle arrest, modulation of the expression of cyclin-dependent kinase inhibitors, and autophagy. J Neurosurg 98(2):378–384
58. Apel A, Herr I, Schwarz H, Rodemann HP, Mayer A (2008) Blocked autophagy sensitizes resistant carcinoma cells to radiation therapy. Cancer Res 68(5):1485–1494
59. Gozuacik D, Kimchi A (2004) Autophagy as a cell death and tumor suppressor mechanism. Oncogene 23(16):2891–2906
60. Lomonaco SL, Finniss S, Xiang C, Decarvalho A, Umansky F, Kalkanis SN, Mikkelsen T, Brodie C (2009) The induction of autophagy by gamma-radiation contributes to the radioresistance of glioma stem cells. Int J Cancer 125(3):717–722
61. Abdouh M, Facchino S, Chatoo W, Balasingam V, Ferreira J, Bernier G (2009) BMI1 sustains human glioblastoma multiforme stem cell renewal. J Neurosci 29(28):8884–8896
62. Facchino S, Abdouh M, Chatoo W, Bernier G (2010) BMI1 confers radioresistance to normal and cancerous neural stem cells through recruitment of the DNA damage response machinery. J Neurosci 30(30):10096–10111
63. Sharpless NE, Ramsey MR, Balasubramanian P, Castrillon DH, DePinho RA (2004) The differential impact of p16(INK4a) or p19(ARF) deficiency on cell growth and tumorigenesis. Oncogene 23(2):379–385
64. Valk-Lingbeek ME, Bruggeman SW, van Lohuizen M (2004) Stem cells and cancer; the polycomb connection. Cell 118(4):409–418
65. Short SC, Giampieri S, Worku M, Alcaide-German M, Sioftanos G, Bourne S, Lio KI, Shaked-Rabi M, Martindale C (2011) Rad51 inhibition is an effective means of targeting DNA repair in glioma models and $CD133^{+}$ tumor-derived cells. Neuro Oncol 13(5):487–499
66. Squatrito M, Brennan CW, Helmy K, Huse JT, Petrini JH, Holland EC (2010) Loss of ATM/Chk2/p53 pathway components accelerates tumor development and contributes to radiation resistance in gliomas. Cancer Cell 18(6):619–629
67. Cheng L, Wu Q, Huang Z, Guryanova OA, Huang Q, Shou W, Rich JN, Bao S (2011) L1CAM regulates DNA damage checkpoint response of glioblastoma stem cells through NBS1. EMBO J 30(5):800–813
68. Raveh S, Gavert N, Ben-Ze'ev A (2009) L1 cell adhesion molecule (L1CAM) in invasive tumors. Cancer Lett 282(2):137–145
69. Siesser PF, Maness PF (2009) L1 cell adhesion molecules as regulators of tumor cell invasiveness. Cell Adh Migr 3(3):275–277
70. Sherman MH, Bassing CH, Teitell MA (2011) Regulation of cell differentiation by the DNA damage response. Trends Cell Biol 21(5):312–319

Chapter 6
DNA Repair Mechanisms in Breast Cancer Stem Cells

Hong Yin and Jonathan Glass

Abstract Breast cancer stem cells (BCSC) are a small subset in heterogeneous breast cancer cell populations and are responsible for breast cancer initiation. BCSC have stem cell properties. The maintenance and propaganda of BCSC is controlled by an intrinsic stem cell signaling network and regulated by extrinsic environmental factors. BCSC are resistant to chemotherapy and radiation therapy. The therapeutic resistance of BCSC derives from multiple mechanisms. In this chapter, we discuss the potential mechanisms for enhanced DNA repair of BCSC in the response to DNA damage.

Keywords Breast cancer stem cells · Therapeutic resistance · DNA damage · DNA repair

Abbreviations

ADCC	Antibody-dependent cell-mediated cytotoxicity
ALDH	Aldehyde dehydrogenase
ATM	Ataxia telangiectasia mutated
ATR	Ataxia telangiectasia and Rad3 related
BER	Base excision repair
BCSC	Breast cancer stem cells
BRCA 1	Breast cancer type 1 susceptibility protein
Cdc-25s	Cell division cycle 25 homologs
Chk1	CHK1 checkpoint homolog (S. pombe)
Chk2	CHK2 checkpoint homolog (S. pombe)
CXCR4	C-X-C chemokine receptor type 4
DNA-PK	DNA-dependent protein kinase catalytic subunit
DSB	Double-strand DNA breaks
γ-H2AX	Phosphorylated histone 2AX
E2A	Transcription factor 3 (E2A immunoglobulin enhancer binding factors E12/E47)

H. Yin (✉) · J. Glass
Department of Medicine and Feist-Weiller Cancer Center, Louisiana State University Health Sciences Center in Shreveport, Shreveport, LA 71130, USA
e-mail: hyin@lsuhsc.edu

L. A. Mathews et al. (eds.), *DNA Repair of Cancer Stem Cells*,
DOI 10.1007/978-94-007-4590-2_6,

E2F	Transcription activator that binds DNA through the E2 recognition site
ES cells	Embryonic stem cells
ESA	Epithelial specific antigen
Her2	Human epidermal growth factor receptor 2
HIF	Hypoxia-inducible factor
IL-6	Interleukin 6
IL-8	Interleukin 8
MDC1	Mediator of DNA damage checkpoint 1
MLH1	MutL homolog 1, colon cancer, nonpolyposis type 2
MSH2	MutS homolog 2, colon cancer, nonpolyposis type 1
NHEJ	Non-homologous end joining
NMEC	Normal mammary epithelial cells
Oct-4	Octamer-binding transcription factor 4
PCNA	Proliferation cell nuclear antigen
PROCR	Endothelial protein C receptor
PTEN	Phosphatase and tensin homolog
PUMA	BCL2 binding component 3
RAD51	RAD51 homolog (RecA homolog, E. coli)
RPA-70	Replication protein A1 (70Kd)
SDF	Stromal-derived-factor
TCF	T-cell specific, HMG-box
TGF	Transforming growth factor
TWIST	Twist-related protein as a basic helix-loop-helix transcription factor
Wnt	Wingless/int
NOXA	Phorbol-12-myristate-13-acetate-induced protein 1.

6.1 Breast Cancer Stem Cells: Identification, Maintenance, Enrichment, and Clinical Significance

6.1.1 Identification of Breast Cancer Stem Cells

Breast cancer stem cells (BCSC) are a small subset in heterogeneous breast cancer cell populations. The first evidence to support the existence of BCSC came from a study by Al-Haij et al., where they identified and implanted putative cancer stem cells from a human primary breast cancer [1]. As few as 100 breast cancer cells with $CD44^{+}/CD24^{-}$ Lineage^{-} phenotype were able to form tumors in mice, whereas tens of thousands of cells with alternative phenotypes failed to form tumors, indicating that $CD44^{+}/CD24^{-}$ Lineage^{-} breast cancer cells have a self-renewal property and are cancer stem-like cells. In 2005, Ponti et al. reported that $CD44^{+}/CD24^{-}$ breast tumor cells, isolated from sphere cultures of MCF7 cells, were increased up to 1,000-fold in tumor-initiating capacity compared to parental MCF7 cells [2]. In general,

Table 6.1 Summary of several surface markers and functional assays developed for the isolation of BCSC

Cell sources	Isolating methods	Self-renewal determination	Reference
Primary tumor and cell lines	CD44/CD24/ESA	Sphere formation tumorigenicity	[1, 2]
Primary tumor and cell line	ALDH	Sphere formation tumorigenicity	[3–5]
Primary tumor	CD44/CD24/ALDH	Tumorigenicity	[3, 5]
Normal mammary cells and tumor cell lines	PHK626	Sphere formation tumorigenicity	[6]
Cell lines	26S proteasome activity	Sphere formation	[7, 8]
Cell lines	OCT4 promoter activity	Sphere formation tumorigenicity	[9]
Primary tumor and cell lines	Dye efflux	Side population tumorigenicity	[10–12]

the BCSC subset is characterized by (1) self-renewal capacity, (2) symmetric and asymmetric division, (3) enhanced resistance to chemotherapy and radiation therapy, (4) expression of embryonic stem cell genes and enhanced signaling of pathways such as Notch, Wnt/β-catenin, and/or Hedgehog; (5) relatively slow division or quiescence in the cell cycle, and (6) expression specific surface markers. According to the features of BCSC, several surface markers and functional assays have been developed for the isolation of BCSC as summarized in Table 6.1.

Despite several reports using the $CD44^{+}/CD24^{-}$ surface marker phenotype to identify BCSC, some researchers have raised concerns over the identification of BCSC with these cell surface markers. First, application of markers and functional assays listed above gave only an enriched, not a pure BCSC subset. Currently, no single protocol or even combined protocol is guaranteed to obtain an absolutely pure BCSC subset. Second, one set of markers could be applied well for one type of breast cancer cell and applied poorly for another type. For example, ER and Her2 negative HCC38 cells and luminal/$her2^{+}$ SK-BR-3 cells showed that greater than 90 % of cells were ALDH positive, but no cell was $CD44^{+}/CD24^{-}$. In contrast, triple negative MDA-MB-231 and MDA-MB-436 cells showed 70–85 % positivity for $CD44^{+}/CD24^{-}$ cells, but were less than 5 % positive by ALDH function assay [13, 14]. The MDA-MB-231 and MBA-MB-361 cell lines have a much higher percentage of $CD44^{+}/CD24^{-/low}$ cells than the SK-BR-3 cell line, but their colony-forming efficiency in soft agar was less than SK-BR-3 cells [15]. In a separate investigation, it was found that $PROCR^{+}/ESA^{+}$ MDA-MB-231 and MBA-MB-361 cells had a twofold increase in colony forming efficiency compared to bulk cells. Therefore, identification of BCSC by one set of markers or one functional assay is not enough. The proper identification procedure should be test for samples from different sources. Recent investigations of ALDH expression and CD44/CD24 distribution in clinical samples discovered that $ALDH^{+}/CD44^{+}/CD24^{-/low}$ cells were enriched in triple-negative or basal-like subtypes of breast cancer, indicating that either CD44/CD24/ALDH could only identify the BCSC in specific subtypes of breast cancer or that other subtypes of breast cancer lack BCSC [16–20].

6.1.2 The Origin of BCSC

Breast cancer is a heterogeneous disease. To explain the heterogeneity, two models have been proposed: the cancer stem cell and the clonal evolution models [21–23]. The cancer stem cell model suggests that the cancer originates from cancer stem cells, which divide into non-stem cells and stem cells. The non-stem cells proliferate and constitute the bulk of tumor volume, whereas cancer stem cells maintain their self-renewal capacity and contribute to clone formation. This model for breast cancer is currently supported by evidence that few stem-like cancer cells can initiate tumors in mice, and a single stem-like cell can form clonal, non-adherent mammospheres [24]. Unlike normal breast stem cells, BCSC are neither able to differentiate to normal mammary epithelial cells nor give rise to homogenous tumor cells. The heterogeneity of breast cancer may be caused partially by varieties of genetic defects or mutations among breast cancer cells. The clonal evolution model states that a mutant cancer cell will obtain growth preference and expand through clonal growth to form a tumor. The mutant clone should have a homogeneous cell population at least for a period of time. However, this model does not explain the heterogeneous nature of breast tumors. In fact, due to accumulated mutations and genomic instability, breast cancer cells are able to lose control of differentiation direction and to create new stem-like cancer cells through dedifferentiation. Well demonstrated examples of the dedifferentiation process are that normal and neoplastic non-stem cells can spontaneously convert to cells with a stem-like property and that epithelial-like cancer cells can transform to mesenchymal-like cancer cells in breast cancer cells [25–28]. Therefore, the BCSC could originate from aberrant or mutant normal breast stem cells or dedifferentiated breast cancer cells. In practice, BCSC isolated by current protocols could be an enriched mix population of stem-like and progenitor cells. BCSC from different sources (cell lines and patient samples) should demonstrate a variety of stemness.

6.1.3 Maintenance of Self-Renewal in BCSC

Factors that determine the maintenance and expansion of BCSC subset are complicated. According to the current understanding of embryonic and adult stem cells, some intrinsic and microenvironmental factors favor the maintenance and expansion of BCSC subset. Intrinsic factors include the increased expression of stem cell gene such as Oct-4 and the enhanced activation of Notch, Wnt/β-catenin, Hedgehog, polycomb signaling pathways [29–33]. The microenvironmental factors are oxygen levels and other factors, such as TGF-β, insulin growth factor, IL-6, IL-8, and SDF1/CXCR4 [14, 34–41].

6.2 Treatment Resistance and DNA Repair in BCSC

6.2.1 Treatment Resistance in BCSC

In addition to self-renewal, one significant characteristic of cancer stem cells is their relative resistance to treatment. The treatment resistance of cancer stem cells is dominantly reflected in the administration of chemotherapeutic drugs and radiation. The characteristic mechanisms of chemo- and radiation therapy are their action on DNA molecules, and DNA replicated events of tumor cells, which induce varied types of DNA damage and corresponding DNA damage responses. Unlike cytostatic drugs, chemotherapeutic drugs and radiation attempt to eliminate tumor cells. Types of chemotherapy drugs and radiation, DNA damage, and potential DNA repair mechanisms are represented in Table 6.2.

In breast cancer, the resistance of BCSC to treatment has been investigated experimentally. A review paper by Lacerda et al. summarized the most current original experiments on the resistance of BCSC to chemotherapy drugs and radiation [62]. In this summary, six of fifteen experiments were carried out with irradiation. Eight experiments were conducted with chemotherapeutic drugs including paclitaxel, doxorubicin, cisplatin, epirubicin, 5-fluorouracil, methotrexate, and/or cyclophosphamide. One experiment used herceptin plus NK cell-mediated ADCC killing. Data showed that the administration of chemotherapeutic drugs and radiation to breast cancer cells enriched for BCSC. In clinical studies from breast cancer core biopsies, Tanie et al. reported that ALDH positive tumor cells were significantly increased by chemotherapy (paclitaxel (80 mg/m^2/wk followed by four cycles of 5-fluorouracil, 500 mg/m^2, epirubicin 75 mg/m^2, and cyclophosphamide 500 mg/m^2 every 3 weeks) [63]. However, $CD44^+/CD24^{-/low-}$ cells did not change between and after treatment. Li et al. revealed that $CD44^+/CD24^{-/low}$ cells were increased after treatment of Her2-negative breast cancer with chemotherapeutic drugs (docetaxel or doxorubicin and cyclophosphamide) for 12 weeks [64]. The treatment of Her2-positive breast cancer with lapatinib slightly decreased the $CD44^+/CD24^{-low}$ cells. Later work by the same group discovered that $CD44^+/CD24^{-/low}$ mammospheres and Claudin-low signatures were more pronounced in tumor tissue remaining after either endocrine therapy (letrozole) or chemotherapy (docetaxel) [65]. The clinical evidence above supports the treatment resistance of BCSC. These data suggest that breast cancer with a $CD44^+/CD24^{-/low}$ phenotype has a poor prognosis [66].

Drug and radiation resistance poses a significant challenge to the treatment of breast cancer. In fact, BCSC were not only found in primary breast cancer, but could also be adopted more in metastatic or recurrent breast cancer, suggesting that more BCSC occur in later-stages. Although results from experimental investigations demonstrated that BCSC are more resistant to chemotherapy and radiation than non-BCSC, more clinical evidence is needed to determine the role of BCSC in failed treatment with chemotherapy and radiation. Currently, there are more review papers than original works on the treatment resistance of BCSC among the literature.

Table 6.2 Types of chemotherapy drugs and radiation, DNA damage, and potential DNA repair mechanisms

Classification	Drugs	DNA damage type	DNA repair	References
Alkylating agents	Cyclophosphamide Ifosfamide Busulfan Dacarbazine Cisplatin Carboplatin Mitomycin C	Cross-linking Base modification	Homologous recombination Nucleotide excision repair Base excision repair	[42–45]
Antimetabolites	5-fluorouracil Gemcitabine Methotrexate Cytarabine	Base modification	Nucleotide excision repair Base excision repair Alkyltransferas	[46–48]
Mitotic inhibitors	Paclitaxel Vinblastine Estramustine Docetaxel	Uncoupling between DNA replica and mitosis Aberrant cell division (mitotic catastrophe)	Check point response	[49, 50]
Topoisomerase inhibitors	Topotecan Irinotecan Etoposide Camptothecin	Single strand DNA break and double strand DNA break	Homologous recombination Nucleotide excision repair Base excision repair	[51–55]
Radiation	X-ray Gamma ray	Double strand DNA break Single strand DNA break Base modification	Non-homologous end joining Homologous recombination Nucleotide excision repair Base excision repair	[56–60]
Replication inhibitors	Hydroxyurea	Stalled replication fork (double strand DNA break)	Homologous recombination	[61]

6.2.2 *Biological Features of BCSC for the Treatment Resistance*

Why do BCSC exhibit more treatment resistance than do non-BCSC? The question is still being investigated. Currently, some intrinsic properties of BCSC could contribute to their treatment resistance, mainly: (1) The quiescent phase of BCSC. Cells in this phase are insensitive to DNA attack by chemotherapeutic drugs and radiation. (2) The expression of ABC proteins, such as ABCG2 in the cell membrane as efflux pumps to exclude chemotherapy drugs. The cell population with this property has been classified as the side population. (3) The low level of reactive oxygen species (ROS). When compared to non-BCSC, irradiated BCSC generated a low level of ROS and exhibits a high capacity to clean ROS. The deletion of GSH resensitized BCSC to radiation [67]. (4) Active survival pathways in BCSC. These survival pathways include antiapoptosis, NF-kb, and PI3-Akt pathway and play a role in treatment resistance of BCSC. (5) High ALDH activity in BCSC. High ALDH increases the metabolism of chemotherapeutic drugs such as cyclophosphamide [68]. (6) The enhanced DNA repair capacity of BCSC compared to non-BCSC. (7) Micro-environmental factors such as hypoxia, cytokines, and angiogenic factors. These factors affect the response of BCSC to treatment. All of these special features of BCSC determine the cell fate after chemo- or radiation therapy and contribute to treatment resistance.

6.3 DNA Repair Mechanisms of BCSC

6.3.1 Homologous Recombination Plays Roles in BCSC

Although there is literature about treatment resistance of cancer stem cells, the understanding of the mechanism(s) of resistance in cancer stem cells is still lacking, especially in regards to DNA repair in cancer stem cells. Most of the available information comes from radiation-based studies. For instance of breast cancer, enhanced DNA repair raised radiation resistance of BCSC. Early investigations of mesenchymal stem cells in response to radiation showed enhanced DNA repair after radiation [69]. The study by Chen et al. compared the post-radiation survival curves among breast cancer HCC1937 cells, human mesenchymal stem cells, and human lung cancer A549 cells. The human mesenchymal stem cells and human lung cancer A549 cells were relatively radiation resistant. Analysis of cell cycles showed that radiation sensitive HCC1937 cells exhibited a significant G2-M accumulation and a sub-G1 increase suggesting apoptosis. Further comparative analysis of DNA repair response found that radiation resistant cells showed enhanced DNA repair response. Blockage of ATM activation after radiation with wortmannin and caffeine induce radiosensitization of mesenchymal stem cells. The results above suggested that enhanced repair of double strand DNA breaks is a function of the radiation resistance of stem cells.

As previously reported, $CD44^{+}/CD24^{-/low}$ breast cancer cells isolated from MCF7, MBA-MD-231 cells, and primary cultured cells from benign and malignant breast tumor are radiation resistant [70]. The increased resistance is correlated to increased activation of ATM signaling. $CD44^{+}/CD24^{-/low-}$ cells showed increased, and persistent phosphorylation at S1981 compared to control cells. The downstream target of ATM also showed increased phosphorylation to some extent. The application of ATM inhibitor KU55933 almost totally abolished the radiation resistance of $CD44^{+}/CD24^{-/low}$ cells. These results suggested an increased homologous recombination repair in BCSC. However, in this study, no increased non-homologous end joining was found in $CD44^{+}/CD24^{-/low}$ cells with *in vivo* and *in vitro* ligation analysis. The expression of components of non-homologous end joining proteins did not show a difference between BCSC and non-BCSC.

Using mammosphere and monolayer culture of MCF7 cells, Karimi-Busheri et al. analyzed single and double stranded DNA damage and carried out a comprehensive comparison of the DNA damage repair and cell cycle check point proteins of the two cell types [71]. Enhanced survival of MCF-7 mammosphere cells in response to ionizing radiation was accompanied by low ROS generation, increased repair of single strand DNA breaks, and highly expressed nucleotide and base repair protein APE1. In addition, there was a significant reduction of the senescence in mammosphere culture of MCF7 cells, which can be attributed to elevated telomerase activity, as well as reduced p21 expression and retinoblastoma protein phosphorylation. However, in their data, enhanced phosphorylation of ATM and its downstream target p53 was apparently seen.

The cell cycle is a critical determinant factor for radiation sensitivity. Cells in M phase are most sensitive to radiation and cells in late S phase are most resistant [72, 73], and in S phase, DNA repair is most active [74]. DNA damage caused by radiation exhibited predominately as double stranded DNA breaks (DSB). DSB occurring in the G1 phase are mainly repaired by non homologous end joining (NHEJ), while DSB formed in S-G2 phase are fixed by both NHEJ and homologous recombination [75, 76]. Theoretically, most of the cancer stem cells should be in quiescent/G0 status [77] and cells in the G0 phase can be separated from G1 by the low RNA amount in a flow cytometry analysis [8]. Double staining with Hoechst 33342 (DNA) and pyronin Y (RNA) demonstrated that a high percentage of cells from the BCSC population were in the G0 phase compared to non-BCSC (25 vs. 0.59 %). Interestingly, that portion of BCSC in the G0 phase was reduced, and these cells were repopulated into the G2 phase after fractional radiation. In contrast, non-BCSC in the G0 phase were increased after radiation. Analysis of β-galactosidase activity showed that 58 % of non-BCSC were X-gal positive and only 6.4 % of BCSC were positive after fractional radiation, suggesting a differential senescence between BCSC and non-BCSC. More recently, a similar phenomenon was seen in BCSC ($CD24^+/ESA^+$) isolated from the breast cancer cell line MDA-MB-231. After radiation, a larger portion of BCSC were in the S-G2 phase compared to unsorted cells [78]. The S-G2 arrest in BCSC was further demonstrated by the decreased expression of cyclin D and E and increased expression of the proliferation cell nuclear antigen (PCNA). The S-G2 arrest in cell cycle is well correlated with more RAD51 foci and fewer γ-H2AX foci seen in BCSC than unsorted cells. Administration of PCI 124781, an inhibitor of homologous recombination that blocks the formation of RAD51 foci, inhibited homologous recombination, and decreased survival of BCSC. In contrast, PC1 124781 had no effect on the survival of unsorted cells. These results above suggested that homologous recombination could be a dominant DNA repair mechanism in irradiated BCSC of some origins. However, the same effect was not seen on sorted MDA-MB-468 cells. Interestingly, a similar change of the G2 block has been seen in stem-like cells isolated from the H357 cell line and head and neck squamous cell carcinomas after exposure to neocarzinostatin [79]. The increased G2 block in BCSC could be associated with senescent/apoptotic resistance.

6.3.2 Over-Active Checkpoint Response

DNA damage check points consist of two pathways: the ATM/ATR-Chk1/Chk2-Cdc25s and ATM/ATR-p53 pathways [79, 80]. The former is responsible for a fast, reversible response and the later for a slow, irreversible response. Depending on the type of DNA damage, Chk1 is largely phosphorylated by ATR, to a less extent by ATM and Chk2 is predominately phosphorylated by ATM. Activated Chk1 and Chk2, as transducers, phosphorylate a series of downstream proteins to play roles in cell cycle check points and cell fate decision. The earliest report of active DNA damage check points of cancer stem cells came from Bao et al. work. $CD133^+$ glioma cells showed increased radiation resistance and a low level of apoptosis compared with $CD133^-$ cells [81]. The increased resistance was abolished with a G2 checkpoint kinase inhibitor, debromohymenialdisine. Examination of DNA damage response proteins showed that $CD133^+$ cells preferentially activate the DNA damage checkpoint and repair radiation-induced DNA damage more efficiently than $CD133^-$ cells did. Hyperphosphorylated ATM, Rad17, Chk1, and Chk2 were apparently seen in irradiated $CD133^+$ cells. Noticeably, $CD133^+$ exhibited a basal activation of Rad17, Chk1 and Chk2, while the basal activation was not seen in ATM. This means that activation of checkpoint proteins may be independent of their upstream mediator ATM. Perhaps, $CD133^+$ cells have a constitutional activation of checkpoint proteins. In addition to glioma, enhanced activation of checkpoint proteins was shown in BCSC and cancer stem cells from other tissues. The $ALDH^{high}/CD44^+$ cells isolated from MDA-MB-231 and MDA-MB-468 cells were highly resistant to doxorubicin, paclitaxel, and radiation. In contrast to $ALDH^{low}/CD44^-$ cells, $ALDH^{high}/CD44^+$ cells showed increased basal activity in a series of DNA response proteins including Chk1 and Chk2. However, inhibition of ALDH activity by diethylaminobenzaldehyde reduced the radiation resistance of $ALDH^{high}/CD44^+$ cells, but did not affect the basal activity of Chk1 and Chk2 [5].

6.3.3 PTEN/AKT/WNT/β-Catenin and DNA Damage Response in Normal and Tumorous Mammary Stem Cells

The PTEN/Akt pathway plays a role in the regulation of normal and malignant mammary stem/progenitor cell populations. Normal mammary epithelial cells (NMEC) in mammosphere cultures expressed increasing Ser^{380} phosphorylation of PTEN. The phosphorylated PTEN inhibits its phosphatase activity antagonizing Akt phosphorylation. Knockdown of PTEN increased the phosphorylation of Akt and GSK3-β. The activated PI3/Akt signaling supported the self-renewal function of NMEC. Inhibition of PI3/Akt signaling suppressed mammosphere formation and outgrowth of NMEC in NOD/SCID mice. In mammary stem/progenitor cells, the regulation of self-renewal function by PTEN/ was mediated by β-Catenin activation. Activated Akt was able to phosphorylate and deactivate GSK3-β and directly phosphorylate and activate β-catenin, leading to nuclear translocation. Knockdown of β-catenin reduced the number of mammospheres [82]. Wnt signaling plays a critical role in the development of mammary gland through the canonical β-catenin pathway. Oncogenic evidence from MMTV-Wnt1 mice showed extensive lobuloalveolar hyperplasia and the development of mammary carcinoma [83, 84]. In breast cancer, an active Wnt/β-catenin pathway has been found in basal-like breast cancer [85, 86]. Furthermore, Wnt/β-catenin has been demonstrated to mediate radiation resistance in mouse Sca1 positive mammary epithelial cells. Radiation selectively activated β-catenin in $Sca1^+$ cell but not in $Sca1^-$ cells. Cells with stabilized β-catenin showed a high portion of stem-like cells (side population) after radiation [87]. More recently, Zhang et al. demonstrated that tumor initiating cells (TICs) isolated from p53 null mouse mammary tumors repair DNA damage following *in vivo* ionizing radiation more efficiently than the bulk of the tumor cells. However, the initial responses to DNA damage were similar for TICs and bulk tumor cells. More efficient DNA repair could be achieved via the selective activation of the Akt and Wnt/catenin signaling as demonstrated by the increased phosphorylation of Akt and β-catenin [88]. The mechanisms of radiation resistance in cells with active Wnt/β-catenin signaling are largely unknown. Some evidence from non-cancer stem cells suggests that activation of Wnt/β-catenin signaling promotes DNA damage tolerance. When DNA is damaged, poly (ADP-ribose) polymerase-1 (PARP -1) ribosylates its own auto modification domain. The poly ADP-ribosylated PARP-1 dissociates from its partner, TCF-4. This disassociation allowed Ku70 to interact with TCF-4. Ku70 also regulated the binding of ß-catenin to TCF-4. This regulation controlled the expression of Wnt/β-catenin target genes [89]. However, in some cells, activated β-catenin resulted in p53-independent DNA damage and disrupted the DNA repair process [90]. Mutation in β-catenin and Axin was accompanied by the loss of mismatch repair [91].

6.3.4 Notch Signaling and DNA Damage Responses of BCSC

Notch signaling plays a role in stem cell maintenance. The activation of Notch signaling by the addition of a Notch-activating Delta/Serrate/Lag-2 peptide increased the mammosphere formation by tenfold [92]. In addition, increased expression of Notch 1 has been demonstrated in $CD44^+/CD24^{-/low}$ cells in mammosphere culture compared with cells in monolayer culture [93]. Elevated Notch signaling is implied in BCSC [94]. Although the effect of Notch signaling activation on the treatment resistance of BCSC is little known, there is some evidence showing that active Notch signaling plays a resistant role in drug-induced apoptosis. Treatment of Notch signaling activated MCF10A/RBP-Jk and MCF10A/NICD, with the kinase inhibitor staurosporine and the DNA-damaging reagents melphalan and mitoxantrone did not induce apoptosis. Melphalan and mitoxantrone activated p53 function through JNK signaling, followed by induction of p53-dependent apoptosis. Notch activation inhibited JNK phosphorylation and PUMA and NOXA up-regulation. Therefore, activation of Notch signaling blocked the p53 function [95]. In leukemia, Notch is a potential anti-apoptotic p53 target. The primary CLL cells treated with p53 activator nutlin-3 showed an increase in both p53 and Notch1. Shutdown of p53 expression abrogated induction of Notch1 by nutlin [96]. In brain tumors, inhibition of Notch signaling with γ-secretase inhibitors rendered the $CD133^{+}$ glioma stem cells more sensitive to radiation at clinically relevant doses. $CD133^{-}$ non-stem cells showed no response to these inhibitors. The enhanced radiation resistance by Notch was due to altered DNA damage response, activation of the PI3K/Akt pathway, and up-regulation of the antiapoptotic proteins, Bcl-2 and Mcl-1 [97]. In addition to anti-apoptosis, Notch signaling has extensive cross-talk with some survival and oncogenic pathways, such Wnt, RAS, Her2, TGFβ, PI3K/Akt, p53, NF-κβ, and HIF [94]. The cross-talk networks of Notch signaling have been considered as a target in breast cancer therapy, especially in dealing with BCSC.

6.3.5 Other Factors Affect DNA Damage Response of BCSC

In addition to what we have mentioned above, there are several factors that could affect the DNA damage responses of BCSC. These factors are hypoxia, survivin, an anti-apoptotic protein, and deficient or mutant BRCA1 or 2.

6.3.5.1 Hypoxia

A hypoxic environment for a tumor is a contributive factor to promote BCSC by regulating stem cell genes and pathways, and inducing de-differentiation [34, 35, 98–101]. The development of treatment resistance under hypoxia is mainly mediated through HIF pathways. In response to DNA damage reagents and radiation, hypoxic cells showed a reduced DNA damage and altered DNA repair mechanisms. Several genes, such as RAD51, BRCA1, MSH2, and MLH1, responsible for homologous recombination and mismatch repair were downregulated. In response to etoposide, a topoisomerase poison, tumor cells pre-exposed to hypoxia diminished DNA damage as determined by comet assay, compared with cells under normoxia. DNA repair capacity of cells was not changed apparently by hypoxia [102]. The damaged DNA responds to hypoxic conditions in two ways. In the initial, early response to hypoxia and re-oxygenation, damaged DNA activates ATM/ATR checkpoint signaling to initiate DNA repair. In the following, chronic response, DNA repair is inhibited by the decreased expression of the genes responsible for several DNA repair mechanisms [103]. For example, the DNA mismatch repair response was inhibited under hypoxic conditions due to the decreased expression of two mismatch repair genes, MLH1 and MSH2. Reduced MLH1 expression could be the result of increased deacetylation and di- and tri-methylation of histone H3 (H3K9) [104]. Hypoxia significantly decreased the binding of c-Myc to the promoter of MLH1 and MSH2 genes [105]. In addition to reducing the function of mismatch repair, hypoxia repressed the function of homologous recombination through the down-regulation of BRCA1 and RAD 51 gene expression by increasing the binding of the inhibitory E2F4/p130 complex to BRCA1 and RAD51 promoters [106, 107]. Interestingly, the hypoxia-mediated down-regulation of DNA repair genes was not HIF-dependent. A recent report revealed that activated DNA-PK mediated phosphorylation of p53(ser15) in hypoxic cells and that the phosphorylated p53 then disassociated from the p53-RAP70 complex. The RPA70 enhanced the efficiency of nucleotide excision and NHEJ DNA repair [108]. It seems that the effects of hypoxia on DNA repair are complicated, and the resulting DNA repair depends on the type of damage done to the DNA and the involved DNA repair mechanisms. Hypoxia-induced drug resistance was closely related to the enhanced anti-apoptotic and/or anti-senescent capacity of cells, including breast cancer cells [109, 110]. The inhibition of senescence under hypoxia is mediated through the down-regulation of E2A-p21 by HIF-TWIST in mesenchymal stem cells [111]. It appears that the effects of hypoxia on treatment resistance may depend on the types of treatments and hypoxia. The pattern of hypoxia affects the development of treatment resistance [102]. Two HIF-α genes have been identified as HIF-1α and HIF-2α, both of which are induced in response to hypoxia. With a long term of hypoxia, induction of HIF-1α is repressed, and HIF-2α stably elevated [110, 112, 113]. This switch may be mediated by the altered level of hypoxia-associated protein [114, 115]. Expression of HIF-2α was preferential in neuronal cancer stem cells. Interestingly, high expression of HIF-2α was correlated to long-term survival in breast cancer patients and to a better response to the classic therapy [116]. HIF-2α up-regulated the expression of Amphiregulin and WNT1-inducible signaling pathway protein-2 and in turn, could be associated with a luminal epithelial differentiation of tumor cells.

6.3.5.2 Survivin

Survivin is an anti-apoptotic protein. In the report of Woodward et al., Sca1$^+$ cells showed an increased survivin level in response to radiation compared with a matched control [87]. The elevated survivin could be regulated by Wnt/β-catenin signaling. In breast cancer, high transcription of Notch-1 and BIRC5 (survivin) is linked to basal-like phenotype cancer which is enriched in BCSC [117]. In response to radiation, nuclear accumulation of survivin was demonstrated [118]. The nuclear survivin interacted with KU70, MDC1, DNA-PKcs, and γ-H2AX. Survivin knockdown resulted in a hampered phosphorylation of DNA-PKcs and an impaired DNA repair capacity. The comparison of responses to radiation among human ES cells, normal somatic lung fibroblasts, and breast cancer cell line MDA-MB-231 cells revealed that ES cell exhibited G2 arrest and differential checkpoint responses compared with control fibroblasts. Elevated survivin expression was found in ES and MDA-MB-231 cells. However, the deletion of survivin in ES cells did not have a significant impact on post-radiation survival of ES cells at least in the first 48 hours after radiation, suggesting a potentially different function of survivin in ES cells [69].

6.3.5.3 Mutation in BRCA1 Gene

Women with mutations in the *BRCA1* gene have a 70–80 % chance to develop breast cancer. Most of these tumors are a basal-like phenotype, characterized by the expression of myoepithelial markers, but lack expression of ER, PR, and Her2 receptors. Therefore, these tumors are enriched in BCSC [119–121]. BRCA1 is important for breast cell differentiation and is a component of DNA repair through homologous recombination. The loss of function of mutant BRCA1 causes blocked differentiation and aberrant DNA repair in response to genetic stress and in turn, leads to genetic instability [122–125]. Theoretically, cancer cells with BRCA1 mutant should be more sensitive to the DNA damage reagents or drugs and expect a better response to treatment. However, this expectation may become compromised due to the unbalanced compensations by other DNA repair mechanisms. Currently, PARP inhibitors, which function in several DNA repair mechanisms such as BER and NHEJ, have been used in or have completed clinical trials [126, 127].

6.4 Conclusions

In conclusion, the available data about DNA repair mechanism(s) utilized by BCSC are less than what we expected and mostly originate from the study of a couple of breast cell lines. The knowledge of the mechanism(s) of DNA repair in BCSC remains unknown. It seems that checkpoint responses and homologous recombination play pivotal roles in the DNA repair of BCSC in response to radiation. The over activation of the DNA repair mechinery in cancer stem cells could be a complex process. For example, constitutive activation of checkpoint proteins may be related to the quiescent status of cells, and whether or not activated checkpoint proteins contribute to DNA repair is currently being examined [128]. The increased activation of survival pathways also attenuates treatment-induced apoptosis and senescence. Similar to the heterogeneity seen in breast cancer, BCSC are a heterogeneous population. BCSC from an individual patient may have variable stemness and BCSC from the same tumor may also exhibit variation in the stemness. The variation in stemness could be based on the genetic variation among cell populations and local microenvironment. Therefore, a difference in DNA repair response may exist in BCSC populations. Paying attention to the personalized treatment of breast cancer may increase the chance of eradicating BCSC and benefit the breast cancer patient's prognosis through better treatment.

References

1. Al-Hajj M, Wicha MS, Benito-Hernandez A, Morrison SJ, Clarke MF (2003) Prospective identification of tumorigenic breast cancer cells. Proc Natl Acad Sci USA 100(7):3983–3988
2. Ponti D, Costa A, Zaffaroni N, Pratesi G, Petrangolini G, Coradini D, Pilotti S, Pierotti MA, Daidone MG (2005) Isolation and in vitro propagation of tumorigenic breast cancer cells with stem/progenitor cell properties. Cancer Res 65(13):5506–5511

3. Ginestier C, Hur MH, Charafe-Jauffret E, Monville F, Dutcher J, Brown M, Jacquemier J, Viens P, Kleer CG, Liu S et al (2007) ALDH1 is a marker of normal and malignant human mammary stem cells and a predictor of poor clinical outcome. Cell Stem Cell 1(5):555–567
4. Morimoto K, Kim SJ, Tanei T, Shimazu K, Tanji Y, Taguchi T, Tamaki Y, Terada N, Noguchi S (2009) Stem cell marker aldehyde dehydrogenase 1-positive breast cancers are characterized by negative estrogen receptor, positive human epidermal growth factor receptor type 2, and high Ki67 expression. Cancer Sci 100(6):1062–1068
5. Croker AK, Goodale D, Chu J, Postenka C, Hedley BD, Hess DA, Allan AL (2009) High aldehyde dehydrogenase and expression of cancer stem cell markers selects for breast cancer cells with enhanced malignant and metastatic ability. J Cell Mol Med 13(8B):2236–2252
6. Pece S, Tosoni D, Confalonieri S, Mazzarol G, Vecchi M, Ronzoni S, Bernard L, Viale G, Pelicci PG, Di Fiore PP (2010) Biological and molecular heterogeneity of breast cancers correlates with their cancer stem cell content. Cell 140(1):62–73
7. Vlashi E, Kim K, Lagadec C, Donna LD, McDonald JT, Eghbali M, Sayre JW, Stefani E, McBride W, Pajonk F (2009) In vivo imaging, tracking, and targeting of cancer stem cells. J Natl Cancer Inst 101(5):350–359
8. Lagadec C, Vlashi E, Della Donna L, Meng Y, Dekmezian C, Kim K, Pajonk F (2010) Survival and self-renewing capacity of breast cancer initiating cells during fractionated radiation treatment. Breast Cancer Res 12(1):R13
9. Sajithlal GB, Rothermund K, Zhang F, Dabbs DJ, Latimer JJ, Grant SG, Prochownik EV (2010) Permanently blocked stem cells derived from breast cancer cell lines. Stem Cells 28(6):1008–1018
10. Clarke RB (2005) Isolation and characterization of human mammary stem cells. Cell Prolif 38(6):375–386
11. Engelmann K, Shen H, Finn OJ (2008) MCF7 side population cells with characteristics of cancer stem/progenitor cells express the tumor antigen MUC1. Cancer Res 68(7):2419–2426
12. Patrawala L, Calhoun T, Schneider-Broussard R, Zhou J, Claypool K, Tang DG (2005) Side population is enriched in tumorigenic, stem-like cancer cells, whereas $ABCG2^+$ and $ABCG2^-$ cancer cells are similarly tumorigenic. Cancer Res 65(14):6207–6219
13. Sheridan C, Kishimoto H, Fuchs RK, Mehrotra S, Bhat-Nakshatri P, Turner CH, Goulet R, Jr., Badve S, Nakshatri H (2006) $CD44^+/CD24^-$ breast cancer cells exhibit enhanced invasive properties: an early step necessary for metastasis. Breast Cancer Res 8(5):R59
14. Charafe-Jauffret E, Ginestier C, Iovino F, Wicinski J, Cervera N, Finetti P, Hur MH, Diebel ME, Monville F, Dutcher J et al (2009) Breast cancer cell lines contain functional cancer stem cells with metastatic capacity and a distinct molecular signature. Cancer Res 69(4):1302–1313
15. Hwang-Verslues WW, Kuo WH, Chang PH, Pan CC, Wang HH, Tsai ST, Jeng YM, Shew JY, Kung JT, Chen CH et al (2009) Multiple lineages of human breast cancer stem/progenitor cells identified by profiling with stem cell markers. PLoS One 4(12):e8377
16. Reuben JM, Lee BN, Gao H, Cohen EN, Mego M, Giordano A, Wang X, Lodhi A, Krishnamurthy S, Hortobagyi GN et al (2011) Primary breast cancer patients with high risk clinicopathologic features have high percentages of bone marrow epithelial cells with ALDH activity and CD44CD24lo cancer stem cell phenotype. Eur J Cancer 47(10):1527–1536
17. Ricardo S, Vieira AF, Gerhard R, Leitao D, Pinto R, Cameselle-Teijeiro JF, Milanezi F, Schmitt F, Paredes J (2011) Breast cancer stem cell markers CD44, CD24 and ALDH1: expression distribution within intrinsic molecular subtype. J Clin Pathol 64(11):937–946
18. Dontu G (2008) Breast cancer stem cell markers—the rocky road to clinical applications. Breast Cancer Res 10(5):110
19. Kai K, Arima Y, Kamiya T, Saya H (2010) Breast cancer stem cells. Breast Cancer 17(2):80–85
20. Nakshatri H, Srour EF, Badve S (2009) Breast cancer stem cells and intrinsic subtypes: controversies rage on. Curr Stem Cell Res Ther 4(1):50–60
21. Campbell LL, Polyak K (2007) Breast tumor heterogeneity: cancer stem cells or clonal evolution? Cell Cycle 6(19):2332–2338
22. O'Brien CA, Kreso A, Jamieson CH (2010) Cancer stem cells and self-renewal. Clin Cancer Res 16(12):3113–3120

23. Takebe N, Ivy SP (2010) Controversies in cancer stem cells: targeting embryonic signaling pathways. Clin Cancer Res 16(12):3106–3112
24. Dontu G, Al-Hajj M, Abdallah WM, Clarke MF, Wicha MS (2003) Stem cells in normal breast development and breast cancer. Cell proliferation 36(Suppl 1):59–72
25. Chaffer CL, Brueckmann I, Scheel C, Kaestli AJ, Wiggins PA, Rodrigues LO, Brooks M, Reinhardt F, Su Y, Polyak K et al (2011) Normal and neoplastic nonstem cells can spontaneously convert to a stem-like state. Proc Natl Acad Sci USA 108(19):7950–7955
26. Mani SA, Guo W, Liao MJ, Eaton EN, Ayyanan A, Zhou AY, Brooks M, Reinhard F, Zhang CC, Shipitsin M et al (2008) The epithelial-mesenchymal transition generates cells with properties of stem cells. Cell 133(4):704–715
27. Takebe N, Warren RQ, Ivy SP (2011) Breast cancer growth and metastasis: interplay between cancer stem cells, embryonic signaling pathways and epithelial-to-mesenchymal transition. Breast Cancer Res 13(3):211
28. Hollier BG, Evans K, Mani SA (2009) The epithelial-to-mesenchymal transition and cancer stem cells: a coalition against cancer therapies. J Mammary Gland Biol Neoplasia 14(1):29–43
29. Liu CG, Lu Y, Wang BB, Zhang YJ, Zhang RS, Chen B, Xu H, Jin F, Lu P (2011) Clinical implications of stem cell gene Oct-4 expression in breast cancer. Ann Surg 253(6):1165–1171
30. Korkaya H, Liu S, Wicha MS (2011) Breast cancer stem cells, cytokine networks, and the tumor microenvironment. J Clin Invest 121(10):3804–3809
31. Woodward WA, Chen MS, Behbod F, Rosen JM (2005) On mammary stem cells. J Cell Sci 118(Pt 16):3585–3594
32. Chase A, Cross NC (2011) Aberrations of EZH2 in cancer. Clin Cancer Res 17(9):2613–2618
33. Chang CJ, Yang JY, Xia W, Chen CT, Xie X, Chao CH, Woodward WA, Hsu JM, Hortobagyi GN, Hung MC (2011) EZH2 promotes expansion of breast tumor initiating cells through activation of RAF1-beta-catenin signaling. Cancer Cell 19(1):86–100
34. Oliveira-Costa JP, Zanetti JS, Silveira GG, Soave DF, Oliveira LR, Zorgetto VA, Soares FA, Zucoloto S, Ribeiro-Silva A (2011) Differential expression of HIF-1alpha in $CD44^{+}$ $CD24^{-/low}$ breast ductal carcinomas. Diagnostic Pathology 6:73
35. Louie E, Nik S, Chen JS, Schmidt M, Song B, Pacson C, Chen XF, Park S, Ju J, Chen EI (2010) Identification of a stem-like cell population by exposing metastatic breast cancer cell lines to repetitive cycles of hypoxia and reoxygenation. Breast Cancer Res 12(6):R94
36. Watabe T, Miyazono K (2009) Roles of TGF-beta family signaling in stem cell renewal and differentiation. Cell Res 19(1):103–115
37. Tan AR, Alexe G, Reiss M (2009) Transforming growth factor-beta signaling: emerging stem cell target in metastatic breast cancer? Breast Cancer Res Treat 115(3):453–495
38. Wang Y, Yu Y, Tsuyada A, Ren X, Wu X, Stubblefield K, Rankin-Gee EK, Wang SE (2011) Transforming growth factor-beta regulates the sphere-initiating stem cell-like feature in breast cancer through miRNA-181 and ATM. Oncogene 30(12):1470–1480
39. Savarese TM, Strohsnitter WC, Low HP, Liu Q, Baik I, Okulicz W, Chelmow DP, Lagiou P, Quesenberry PJ, Noller KL et al (2007) Correlation of umbilical cord blood hormones and growth factors with stem cell potential: implications for the prenatal origin of breast cancer hypothesis. Breast Cancer Res: BCR 9(3):R29
40. Iliopoulos D, Hirsch HA, Wang G, Struhl K (2011) Inducible formation of breast cancer stem cells and their dynamic equilibrium with non-stem cancer cells via IL6 secretion. Proc Natl Acad Sci USA 108(4):1397–1402
41. Huang M, Li Y, Zhang H, Nan F (2010) Breast cancer stromal fibroblasts promote the generation of $CD44^{+}CD24^{-}$ cells through SDF-1/CXCR4 interaction. J Exp Clin Cancer Res CR 29:80
42. Vasquez KM (2010) Targeting and processing of site-specific DNA interstrand crosslinks. Environ Mol Mutagen 51(6):527–539
43. Sarkaria JN, Kitange GJ, James CD, Plummer R, Calvert H, Weller M, Wick W (2008) Mechanisms of chemoresistance to alkylating agents in malignant glioma. Clin Cancer Res 14(10):2900–2908

44. Drablos F, Feyzi E, Aas PA, Vaagbo CB, Kavli B, Bratlie MS, Pena-Diaz J, Otterlei M, Slupphaug G, Krokan HE (2004) Alkylation damage in DNA and RNA—repair mechanisms and medical significance. DNA repair 3(11):1389–1407
45. Kaina B, Christmann M (2002) DNA repair in resistance to alkylating anticancer drugs. Int J Clin Pharmacol Ther 40(8):354–367
46. Rosell R, Crino L, Danenberg K, Scagliotti G, Bepler G, Taron M, Alberola V, Provencio M, Camps C, De Marinis F et al (2003) Targeted therapy in combination with gemcitabine in non-small cell lung cancer. Semin Oncol 30(4 Suppl 10):19–25
47. Wyatt MD, Wilson DM, 3rd (2009) Participation of DNA repair in the response to 5-fluorouracil. Cell Mol Life Sci 66(5):788–799
48. Mounetou E, Debiton E, Buchdahl C, Gardette D, Gramain JC, Maurizis JC, Veyre A, Madelmont JC (1997) O6-(alkyl/aralkyl)guanosine and 2Ê¹-deoxyguanosine derivatives: synthesis and ability to enhance chloroethylnitrosourea antitumor action. J Med Chem 40(18):2902–2909
49. Yim EK, Bae JS, Lee SB, Lee KH, Kim CJ, Namkoong SE, Um SJ, Park JS (2004) Proteome analysis of differential protein expression in cervical cancer cells after paclitaxel treatment. Cancer Res Treat 36(6):395–399
50. Yalowich JC (1987) Effects of microtubule inhibitors on etoposide accumulation and DNA damage in human K562 cells in vitro. Cancer Res 47(4):1010–1015
51. Sortibran AN, Tellez MG, Rodriguez-Arnaiz R (2006) Genotoxic profile of inhibitors of topoisomerases I (camptothecin) and II (etoposide) in a mitotic recombination and sex-chromosome loss somatic eye assay of Drosophila melanogaster. Mutat Res 604(1–2):83–90
52. Heisig P (2009) Type II topoisomerases—inhibitors, repair mechanisms and mutations. Mutagenesis 24(6):465–469
53. James E, Waldron-Lynch MG, Saif MW (2009) Prolonged survival in a patient with BRCA2 associated metastatic pancreatic cancer after exposure to camptothecin: a case report and review of literature. Anticancer Drugs 20(7):634–638
54. Malik M, Nitiss JL (2004) DNA repair functions that control sensitivity to topoisomerase-targeting drugs. Eukaryotic Cell 3(1):82–90
55. Miyagawa K (2008) Clinical relevance of the homologous recombination machinery in cancer therapy. Cancer Sci 99(2):187–194
56. Mahaney BL, Meek K, Lees-Miller SP (2009) Repair of ionizing radiation-induced DNA double-strand breaks by non-homologous end-joining. Biochem J 417(3):639–650
57. Tomita M (2010) Involvement of DNA-PK and ATM in radiation- and heat-induced DNA damage recognition and apoptotic cell death. J Radiat Res 51(5):493–501
58. Powell SN, Kachnic LA (2003) Roles of BRCA1 and BRCA2 in homologous recombination, DNA replication fidelity and the cellular response to ionizing radiation. Oncogene 22(37):5784–5791
59. Murray D, Vallee-Lucic L, Rosenberg E, Andersson B (2002) Sensitivity of nucleotide excision repair-deficient human cells to ionizing radiation and cyclophosphamide. Anticancer Res 22(1A):21–26
60. Hafer K, Iwamoto KS, Scuric Z, Schiestl RH (2007) Adaptive response to gamma radiation in mammalian cells proficient and deficient in components of nucleotide excision repair. Radiat Res 168(2):168–174
61. Petermann E, Orta ML, Issaeva N, Schultz N, Helleday T (2010) Hydroxyurea-stalled replication forks become progressively inactivated and require two different RAD51-mediated pathways for restart and repair. Mol Cell 37(4):492–502
62. Lacerda L, Pusztai L, Woodward WA (2010) The role of tumor initiating cells in drug resistance of breast cancer: implications for future therapeutic approaches. Drug Resist Updat 13(4–5):99–108
63. Tanei T, Morimoto K, Shimazu K, Kim SJ, Tanji Y, Taguchi T, Tamaki Y, Noguchi S (2009) Association of breast cancer stem cells identified by aldehyde dehydrogenase 1 expression with resistance to sequential Paclitaxel and epirubicin-based chemotherapy for breast cancers. Clin Cancer Res 15(12):4234–4241

64. Li X, Lewis MT, Huang J, Gutierrez C, Osborne CK, Wu MF, Hilsenbeck SG, Pavlick A, Zhang X, Chamness GC et al (2008) Intrinsic resistance of tumorigenic breast cancer cells to chemotherapy. J Natl Cancer Inst 100(9):672–679
65. Creighton CJ, Li X, Landis M, Dixon JM, Neumeister VM, Sjolund A, Rimm DL, Wong H, Rodriguez A, Herschkowitz JI et al (2009) Residual breast cancers after conventional therapy display mesenchymal as well as tumor-initiating features. Proc Natl Acad Sci USA 106(33):13820–13825
66. Ahmed MA, Aleskandarany MA, Rakha EA, Moustafa RZ, Benhasouna A, Nolan C, Green AR, Ilyas M, Ellis IO (2011) A CD44$^{(-)}$/CD24$^{(+)}$ phenotype is a poor prognostic marker in early invasive breast cancer. Breast Cancer Res Treat. [Epub ahead of print]
67. Diehn M, Cho RW, Lobo NA, Kalisky T, Dorie MJ, Kulp AN, Qian D, Lam JS, Ailles LE, Wong M et al (2009) Association of reactive oxygen species levels and radioresistance in cancer stem cells. Nature 458(7239):780–783
68. Pearce DJ, Taussig D, Simpson C, Allen K, Rohatiner AZ, Lister TA, Bonnet D (2005) Characterization of cells with a high aldehyde dehydrogenase activity from cord blood and acute myeloid leukemia samples. Stem Cells 23(6):752–760
69. Chen MF, Lin CT, Chen WC, Yang CT, Chen CC, Liao SK, Liu JM, Lu CH, Lee KD (2006) The sensitivity of human mesenchymal stem cells to ionizing radiation. Int J Radiat Oncol Biol Phys 66(1):244–253
70. Yin H, Glass J (2011) The phenotypic radiation resistance of CD44^{+}/CD24$^{(-\text{ or low})}$ breast cancer cells is mediated through the enhanced activation of ATM signaling. PLoS One 6(9):e24080
71. Karimi-Busheri F, Rasouli-Nia A, Mackey JR, Weinfeld M (2010) Senescence evasion by MCF-7 human breast tumor-initiating cells. Breast Cancer Res 12(3):R31
72. Pajonk F, Vlashi E, McBride WH (2010) Radiation resistance of cancer stem cells: the 4 R's of radiobiology revisited. Stem Cells 28(4):639–648
73. Pawlik TM, Keyomarsi K (2004) Role of cell cycle in mediating sensitivity to radiotherapy. Int J Radiat Oncol Biol Phys 59(4):928–942
74. Branzei D, Foiani M (2008) Regulation of DNA repair throughout the cell cycle. Nat Rev Mol Cell Biol 9(4):297–308
75. Mao Z, Bozzella M, Seluanov A, Gorbunova V (2008) DNA repair by nonhomologous end joining and homologous recombination during cell cycle in human cells. Cell Cycle 7(18):2902–2906
76. Tomashevski A, Webster DR, Grammas P, Gorospe M, Kruman, II (2010) Cyclin-C-dependent cell-cycle entry is required for activation of non-homologous end joining DNA repair in postmitotic neurons. Cell Death Differ 17(7):1189–1198
77. Moore N, Lyle S (2011) Quiescent, slow-cycling stem cell populations in cancer: a review of the evidence and discussion of significance. J Oncol 2011:11 (Article ID 396076)
78. Al-Assar O, Mantoni T, Lunardi S, Kingham G, Helleday T, Brunner TB (2011) Breast cancer stem-like cells show dominant homologous recombination due to a larger S-G2 fraction. Cancer Biol Ther 11(12):1028–1035
79. Harper LJ, Costea DE, Gammon L, Fazil B, Biddle A, Mackenzie IC (2010) Normal and malignant epithelial cells with stem-like properties have an extended G2 cell cycle phase that is associated with apoptotic resistance. BMC Cancer 10:166
80. Shrivastav M, De Haro LP, Nickoloff JA (2008) Regulation of DNA double-strand break repair pathway choice. Cell Res 18(1):134–147
81. Bao S, Wu Q, McLendon RE, Hao Y, Shi Q, Hjelmeland AB, Dewhirst MW, Bigner DD, Rich JN (2006) Glioma stem cells promote radioresistance by preferential activation of the DNA damage response. Nature 444(7120):756–760
82. Korkaya H, Paulson A, Charafe-Jauffret E, Ginestier C, Brown M, Dutcher J, Clouthier SG, Wicha MS (2009) Regulation of mammary stem/progenitor cells by PTEN/Akt/beta-catenin signaling. PLoS Biology 7(6):e1000121
83. Verani R, Cappuccio I, Spinsanti P, Gradini R, Caruso A, Magnotti MC, Motolese M, Nicoletti F, Melchiorri D (2007) Expression of the Wnt inhibitor Dickkopf-1 is required for the induction

of neural markers in mouse embryonic stem cells differentiating in response to retinoic acid. J Neurochem 100(1):242–250
84. Semenov M, Tamai K, He X (2005) SOST is a ligand for LRP5/LRP6 and a Wnt signaling inhibitor. J Biol Chem 280(29):26770–26775
85. Khramtsov AI, Khramtsova GF, Tretiakova M, Huo D, Olopade OI, Goss KH (2010) Wnt/beta-catenin pathway activation is enriched in basal-like breast cancers and predicts poor outcome. Am J Pathol 176(6):2911–2920
86. Geyer FC, Lacroix-Triki M, Savage K, Arnedos M, Lambros MB, MacKay A, Natrajan R, Reis-Filho JS (2011) Beta-Catenin pathway activation in breast cancer is associated with triple-negative phenotype but not with CTNNB1 mutation. Mod Pathol 24(2):209–231
87. Woodward WA, Chen MS, Behbod F, Alfaro MP, Buchholz TA, Rosen JM (2007) WNT/beta-catenin mediates radiation resistance of mouse mammary progenitor cells. Proc Natl Acad Sci USA 104(2):618–623
88. Zhang M, Behbod F, Atkinson RL, Landis MD, Kittrell F, Edwards D, Medina D, Tsimelzon A, Hilsenbeck S, Green JE et al (2008) Identification of tumor-initiating cells in a p53-null mouse model of breast cancer. Cancer Res 68(12):4674–4682
89. Idogawa M, Masutani M, Shitashige M, Honda K, Tokino T, Shinomura Y, Imai K, Hirohashi S, Yamada T (2007) Ku70 and poly(ADP-ribose) polymerase-1 competitively regulate beta-catenin and T-cell factor-4-mediated gene transactivation: possible linkage of DNA damage recognition and Wnt signaling. Cancer Res 67(3):911–918
90. Xu M, Yu Q, Subrahmanyam R, Difilippantonio MJ, Ried T, Sen JM (2008) Beta-catenin expression results in p53-independent DNA damage and oncogene-induced senescence in prelymphomagenic thymocytes in vivo. Mol Cell Biol 28(5):1713–1723
91. Castiglia D, Bernardini S, Alvino E, Pagani E, De Luca N, Falcinelli S, Pacchiarotti A, Bonmassar E, Zambruno G, D'Atri S (2008) Concomitant activation of Wnt pathway and loss of mismatch repair function in human melanoma. Genes Chromosomes Cancer 47(7):614–624
92. Dontu G, Jackson KW, McNicholas E, Kawamura MJ, Abdallah WM, Wicha MS (2004) Role of Notch signaling in cell-fate determination of human mammary stem/progenitor cells. Breast Cancer Res 6(6):R605–615
93. Phillips TM, McBride WH, Pajonk F (2006) The response of $CD24^{(-/low)}/CD44^{+}$ breast cancer-initiating cells to radiation. J Natl Cancer Inst 98(24):1777–1785
94. Wu F, Stutzman A, Mo YY (2007) Notch signaling and its role in breast cancer. Front Biosci 12:4370–4383
95. Stylianou S, Clarke RB, Brennan K (2006) Aberrant activation of notch signaling in human breast cancer. Cancer Res 66(3):1517–1525
96. Wickremasinghe RG, Prentice AG, Steele AJ (2011) p53 and Notch signaling in chronic lymphocytic leukemia: clues to identifying novel therapeutic strategies. Leukemia 25(9):1400–1407
97. Wang J, Wakeman TP, Lathia JD, Hjelmeland AB, Wang XF, White RR, Rich JN, Sullenger BA (2010) Notch promotes radioresistance of glioma stem cells. Stem Cells 28(1):17–28
98. Helczynska K, Kronblad A, Jogi A, Nilsson E, Beckman S, Landberg G, Pahlman S (2003) Hypoxia promotes a dedifferentiated phenotype in ductal breast carcinoma in situ. Cancer Res 63(7):1441–1444
99. Axelson H, Fredlund E, Ovenberger M, Landberg G, Pahlman S (2005) Hypoxia-induced dedifferentiation of tumor cells—a mechanism behind heterogeneity and aggressiveness of solid tumors. Semin Cell Dev Biol 16(4–5):554–563
100. Holmquist L, Lofstedt T, Pahlman S (2006) Effect of hypoxia on the tumor phenotype: the neuroblastoma and breast cancer models. Adv Exp Med Biol 587:179–193
101. Mathieu J, Zhang Z, Zhou W, Wang AJ, Heddleston JM, Pinna CM, Hubaud A, Stadler B, Choi M, Bar M et al (2011) HIF induces human embryonic stem cell markers in cancer cells. Cancer Res 71(13):4640–4652
102. Sullivan R, Graham CH (2009) Hypoxia prevents etoposide-induced DNA damage in cancer cells through a mechanism involving hypoxia-inducible factor 1. Mol Cancer Ther 8(6):1702–1713

103. Bindra RS, Crosby ME, Glazer PM (2007) Regulation of DNA repair in hypoxic cancer cells. Cancer Metastasis Rev 26(2):249–260
104. Chen H, Yan Y, Davidson TL, Shinkai Y, Costa M (2006) Hypoxic stress induces dimethylated histone H3 lysine 9 through histone methyltransferase G9a in mammalian cells. Cancer Res 66(18):9009–9016
105. Bindra RS, Glazer PM (2007) Co-repression of mismatch repair gene expression by hypoxia in cancer cells: role of the Myc/Max network. Cancer Letters 252(1):93–103
106. Bindra RS, Gibson SL, Meng A, Westermark U, Jasin M, Pierce AJ, Bristow RG, Classon MK, Glazer PM (2005) Hypoxia-induced down-regulation of BRCA1 expression by E2Fs. Cancer Res 65(24):11597–11604
107. Bindra RS, Glazer PM (2007) Repression of RAD51 gene expression by E2F4/p130 complexes in hypoxia. Oncogene 26(14):2048–2057
108. Madan E, Gogna R, Pati U (2012) p53Ser15 Phosphorylation disrupts p53-RPA70 complex and induces RPA70-mediated DNA repair in hypoxia. Biochem J 443:811–820
109. Sullivan R, Pare GC, Frederiksen LJ, Semenza GL, Graham CH (2008) Hypoxia-induced resistance to anticancer drugs is associated with decreased senescence and requires hypoxia-inducible factor-1 activity. Mol Cancer Ther 7(7):1961–1973
110. Rohwer N, Cramer T (2011) Hypoxia-mediated drug resistance: novel insights on the functional interaction of HIFs and cell death pathways. Drug Resist Updat 14(3):191–201
111. Tsai CC, Chen YJ, Yew TL, Chen LL, Wang JY, Chiu CH, Hung SC (2011) Hypoxia inhibits senescence and maintains mesenchymal stem cell properties through down-regulation of E2A-p21 by HIF-TWIST. Blood 117(2):459–469
112. Seton-Rogers S (2011) Hypoxia: HIF switch. Nat Rev Cancer 11(6):391
113. Lin Q, Cong X, Yun Z (2011) Differential hypoxic regulation of hypoxia-inducible factors 1alpha and 2alpha. Mol Cancer Res 9(6):757–765
114. Koh MY, Powis G (2009) HAF: the new player in oxygen-independent HIF-1alpha degradation. Cell Cycle 8(9):1359–1366
115. Koh MY, Lemos R, Jr., Liu X, Powis G (2011) The hypoxia-associated factor switches cells from HIF-1alpha- to HIF-2alpha-dependent signaling promoting stem cell characteristics, aggressive tumor growth and invasion. Cancer Res 71(11):4015–4027
116. Stiehl DP, Bordoli MR, Abreu-Rodriguez I, Wollenick K, Schraml P, Gradin K, Poellinger L, Kristiansen G, Wenger RH (2012) Non-canonical HIF-2alpha function drives autonomous breast cancer cell growth via an AREG-EGFR/ErbB4 autocrine loop. Oncogene 31(18):2283–2297
117. Rennstam K, McMichael N, Berglund P, Honeth G, Hegardt C, Ryden L, Luts L, Bendahl PO, Hedenfalk I (2010) Numb protein expression correlates with a basal-like phenotype and cancer stem cell markers in primary breast cancer. Breast Cancer Res Treat 122(2):315–324
118. Capalbo G, Dittmann K, Weiss C, Reichert S, Hausmann E, Rodel C, Rodel F (2010) Radiation-induced survivin nuclear accumulation is linked to DNA damage repair. Int J Radiat Oncol Biol Phys 77(1):226–234
119. Kakarala M, Wicha MS (2008) Implications of the cancer stem-cell hypothesis for breast cancer prevention and therapy. J Clin Oncol 26(17):2813–2820
120. Molyneux G, Smalley MJ (2011) The cell of origin of BRCA1 mutation-associated breast cancer: a cautionary tale of gene expression profiling. J Mammary Gland Biol Neoplasia 16(1):51–55
121. Liu S, Ginestier C, Charafe-Jauffret E, Foco H, Kleer CG, Merajver SD, Dontu G, Wicha MS (2008) BRCA1 regulates human mammary stem/progenitor cell fate. Proc Natl Acad Sci USA 105(5):1680–1685
122. Al-Wahiby S, Slijepcevic P (2005) Chromosomal aberrations involving telomeres in BRCA1 deficient human and mouse cell lines. Cytogenet Genome Res 109(4):491–496
123. Miyoshi Y, Murase K, Oh K (2008) Basal-like subtype and BRCA1 dysfunction in breast cancers. Int J Clin Oncol 13(5):395–400
124. Furuta S, Jiang X, Gu B, Cheng E, Chen PL, Lee WH (2005) Depletion of BRCA1 impairs differentiation but enhances proliferation of mammary epithelial cells. Proc Natl Acad Sci USA 102(26):9176–9181

125. O'Donovan PJ, Livingston DM (2010) BRCA1 and BRCA2: breast/ovarian cancer susceptibility gene products and participants in DNA double-strand break repair. Carcinogenesis 31(6):961–967
126. Amir E, Seruga B, Serrano R, Ocana A (2010) Targeting DNA repair in breast cancer: a clinical and translational update. Cancer Treat Rev 36(7):557–565
127. Evers B, Drost R, Schut E, de Bruin M, Van Der Burg E, Derksen PW, Holstege H, Liu X, van Drunen E, Beverloo HB et al (2008) Selective inhibition of BRCA2-deficient mammary tumor cell growth by AZD2281 and cisplatin. Clin Cancer Res 14(12):3916–3925
128. Ropolo M, Daga A, Griffero F, Foresta M, Casartelli G, Zunino A, Poggi A, Cappelli E, Zona G, Spaziante R et al (2009) Comparative analysis of DNA repair in stem and nonstem glioma cell cultures. Mol Cancer Res 7(3):383–392

Chapter 7
DNA Repair Mechanisms in Other Cancer Stem Cell Models

Mihoko Kai

Abstract Stem cells are often referred to as the mother of all cells, meaning that they sit at the apex of a cellular hierarchy and, upon differentiation, give rise to all the mature cells of a tissue. DNA damage constantly arises from DNA replication, spontaneous chemical reactions and assaults by external or metabolism-derived agents. Therefore, all living cells must constantly contend with DNA damage. It is particularly crucial for survival of organisms how DNA damage is handled in stem cells, including tissue specific stem cells. While tissue-specific stem cells share the same purpose of maintaining organ functionality, recent studies have shown that the mechanisms of their response to DNA damage, the outcome of their DNA damage response, and the consequence of DNA repair for genomic stability vary greatly between tissues. Striking differences in the outcome of DNA damage response (DDR) have been seen in hematopoietic stem cells from different species and at different developmental stages. Furthermore cell cycle and metabolic states of stem cells seem to affect choices of DNA repair pathways and a choice between cell survival and death.

7.1 Introduction

The cancer stem cell (CSC) model of tumor development and progression states that tumors, like normal adult tissues, contain a subset of cells that both self renew and give rise to differentiated progeny [1]. A number of CSCs have been identified, including leukemia, breast, brain, melanoma, prostate, head and neck squamous cell carcinomas (HNSCC), colon and pancreatic tumors [2–13]. The cellular origin of CSCs remains elusive. However, these CSCs functionally resemble tissue specific stem cells, and share surface markers with adult stem cells. Therefore, it is believed that CSCs are derived from tissue specific stem cells or converted from progenitor cells. Recent studies indicate that CSCs may take advantage of the mechanisms of DNA repair used by tissue specific stem cells to mediate resistance to chemo- and radiotherapy [14]. Understanding of DNA damage response controls in CSCs has

M. Kai (✉)
Department of Radiation Oncology and Molecular Radiation Sciences,
Department of Oncology, The Sidney Kimmel Cancer Center,
Johns Hopkins University School of Medicine CRBII, Room 404. 1550 Orleans St.,
Baltimore, MD 21231, USA
e-mail: mkai2@jhmi.edu

L. A. Mathews et al. (eds.), *DNA Repair of Cancer Stem Cells*,
DOI 10.1007/978-94-007-4590-2_7,

emerged particularly in glioblastoma and breast CSCs (see Chaps. 5 and 6). However, it is still largely unknown how CSCs respond to DNA damage despite its importance in therapies. Unlike tissue specific stem cells, cancer cells are heterogeneous in nature, and often carry mutations in DNA repair and damage response genes. The background mutations might affect the DNA damage response of CSCs. This chapter focuses on various CSCs giving overviews of their DNA damage responses (DDR).

7.2 Cancer Stem Cells in Leukemia

Leukemia was the first disease for which human cancer stem cells, or leukemic stem cells (LSCs) were isolated through the groundbreaking work of Bonnet and Dick [15]. The hematopoietic system is one of the best tissues for investigating cancer stem cells, since the developmental hierarchy of normal blood formation is well defined and distinct subsets of mature and immature hematopoietic cells can be isolated by fluorescence-activated cell sorting (FACS) based on expression of known surface markers [16].

Leukemias often arise due to deregulated hematopoietic stem cell (HSC) functions or acquisition of extended self-renewal capabilities by more mature progenitor cells [14, 17]. Existence of CSCs in several types of human leukemias have been shown [15, 18, 19]. Like hematopoietic stem cells (HSCs), the populations of human LSCs were found to be mainly quiescent [20, 21], and thereby refractory to most of the conventional treatments and as such relapse [16]. LSCs also use other prospective mechanisms of HSCs, including localization to hypoxic niche, and DDR mechanisms, to specifically escape chemo- and radiotherapies that kill the bulk of the tumor cells [14, 22].

Chronic myeloid leukemia (CML) is sustained by a rare population of primitive, quiescent BCR-ABL^{+} cells and represents an excellent example of a malignancy in which CSCs represent the key to disease eradication [23]. In CML, the expanded clone is believed to be initiated in a pluripotent hematopoietic stem cell, by chance occurrence of a rare mutational event, the translocation of t(9;22), giving rise to the Philadelphia (Ph) chromosome and expression of the oncogenic fusion protein tyrosine kinase breakpoint cluster region-abelson (BCR-ABL) [24]. Although the BCR-ABL tyrosine kinase inhibitor imatinib mesylate has revolutionized CML treatment owing to its remarkable clinical efficiency, it does not appear to be fully curative, owing to the likely survival of BCR-ABL expressing HSCs in patients [16, 23, 25].

CML is a two-stage blood disease that can be separated into chronic and acute phases. The patients with chronic phase disease usually respond to treatments with ABL tyrosine kinase inhibitors. However, some patients who respond initially later become resistant. The pleiotropic effect of constitutive BCR-ABL activity seems to cause epigenetic changes [26, 27]. Expression studies demonstrated that BCR-ABL dramatically perturbs the CML transcriptome, resulting in altered expression of genes [28]. The posttranscriptional, translational, and posttranslational effects of high BCR-ABL levels result in the constitutive activation of factors with mitogenic, anti-apoptotic and anti-differentiation activity (e.g. MAPK$^{ERK1/2}$, MYC, JAK2, YES-1,

LYN, hnRNP-E2, MDM2, STAT5, BMI1, and BCL-2) and inhibition of major key regulators of cellular processes, such as those regulated by the tumor suppressors p53, CCAAT/enhancer binding protein-α (C/EBPα), and PP2A [26, 27, 29–31]. Therefore, it is likely that increased BCR-ABL activity promotes clonal evolution and survival of the tumor. Furthermore, there is a direct correlation between the levels of BCR/ABL, the frequency of clinically relevant BCR/ABL mutations and the differentiation arrest of myeloid progenitors [31–34]. It is highly possible that disease progression and maintenance of the CML stem/progenitor cells are caused by the right combination of genetic and epigenetic abnormalities.

The transition from the chronic to the acute stage is poorly understood, but the deregulation of DDR pathways and acquisition of additional chromosomal aberrations and mutations resulting in overall genomic instability in both HSCs and their downstream progeny are believed to play a crucial role in the transition to the malignant state. BCR-ABL-expressing cells have been found to accumulate genetic abnormalities, but the mechanism leading to this genomic instability is controversial [35]. BCR-ABL-transformed cell lines and $CD34^+$ CML cells contain about 2–6 times more reactive oxygen species (ROS) than their normal counterparts, and accumulate 4–8 times more double-strand breaks (DSBs) [36–38]. Unfaithful and/or inefficient DNA repair of ROS-induced oxidized DNA bases and DSBs could lead to a variety of chromosome aberrations [39]. Effects of BCR-ABL on many DNA repair pathways have been described.

7.2.1 Double-Strand Break Repair in BCR-ABL Cells

It is well documented that partial deletions, duplications and translocations are commonly observed in patients with the acute stage disease [40]. These chromosomal aberrations could arise from unfaithful repair of DSBs. Effects of BCR-ABL in DSB repairs have been demonstrated.

Enhanced homologous recombination repair efficiency as well as sister-chromatid exchange frequency in BCR-ABL expressing cells have been shown [41–43]. Indeed the fusion tyrosin kinase-dependent upregulation of Rad51 expression is reported [42]. Furthermore, c-Abl kinase phosphorylates Rad51 in response to ionizing radiation (IR) [44]. Interestingly downregulation of BRCA1, which is a regulator of Rad51, was observed [43].

Non-homologous end joining (NHEJ) repair usually occurs in a cell cycle dependent manner. It is a preferred pathway when cells are in G0/G1 phase of cell cycle. Therefore, the CML cells, which are in a quiescent state, might utilize this pathway preferentially to repair double-strand breaks (DSBs). In fact, NHEJ activity was approximately two-fold higher in BCR-ABL expressing 32Dcl3 cells compared to the parental cells, and four-fold higher in the case of 5′ overhang repair activity. Additionally, more frequent small additions and larger deletions were found in the BCR-ABL expressing cells [41]. Another group confirmed these results in CML patient cells. BCR-ABL-expressing CML patient samples and K562 cells exhibited a three- to five-fold increase in end-ligation efficiency compared to normal CD34$^+$ cells. Larger deletions, 30–400 bp, were observed in the CML cells. It remains controversial whether the activated NHEJ pathway is a cause for genomic instability in CML cells or not. In other studies, no difference in blunt-end repair was seen between K562 myeloid leukemia cells with a p53 mutation and normal human lymphocytes. However, the p53-negative K562 cells induced fewer repair products with 5′ overhangs than normal lymphocytes [45]. Downregulation of DNA-PKcs but not Ku70 and Ku80 were observed by one group [46]. It is not clear whether elevated levels of DNA damage are driving error-prone repair by NHEJ in CML cells or CML cells activate the NHEJ pathway inducing genomic instability.

7.2.2 Other Repairs in BCR-ABL Cells

BCR/ABL oncogenic tyrosine kinase exhibits two complementary roles in cancer development. The first and best-characterized role is stimulation of signaling pathways that eventually induce growth-factor independence and affect the adhesive and invasive capability of leukemia cells. The second is modulation of response to DNA damage rendering cells resistant to genotoxic therapies and causing genomic instability as described above. BCR/ABL-induced genomic instability may lead to mutations and chromosomal translocations frequently observed during the transition from a relatively benign CML chronic phase (CML-CP) to an aggressive blast crisis (CML-BC) [26, 37, 47, 48]. Mechanisms leading to resistance include amplification of the BCR/ABL gene and acquired additional genomic alterations, which are likely to be caused by deregulation of DSB repair pathways as discussed above. Beside these gross chromosomal changes, numerous small mutations are detected in the BCR/ABL gene itself encoding for resistance to imatinib mesylate [37, 49, 50]. ROS induced by BCR/ABL expression and clonal selection during evolution of the disease seems to be a cause of the mutations that are detected in patient cells.

It has also been reported that BCR/ABL inhibits mismatch repair (MMR) leading to accumulation of mutations. Impaired MMR activity is associated with better survival, accumulation of p53 and lack of activation of Caspase 3 after N-methyl-N′-nitro-N-nitrosoguanidine (MNNG) treatment [51]. Microsatellite instability was observed in CML-BC but not in relatively benign CML-CP. This microsatellite instability seems to reflect multiple replication errors due to defective MMR [52].

Connections between nucleotide excision repair (NER) and BCR/ABL have been indicated. Interaction between XPB and p210 BCR/ABL (but not p185 BCR/ABL) has been shown. It was later suggested that NER defect seen in BCR/ABL cells might be a result of BCR/ABL interfering with overall formation of TFIIH complex formation [53–56]. Ectopic expression of p210 BCR/ABL in murine lymphoid cell line inhibits NER activity *in vitro*, promoting hypersensitivity of these cells to ultraviolet (UV) treatment and facilitating a mutator phenotype. However, expression of p210 BCR/ABL in human and murine myeloid cell lines and primary bone marrow cells resulted in the increased NER activity and resistance to UV irradiation [57]. Furthermore, it was shown that stably expressing BCR/ABL human hematopoietic cell lines as well as fibroblast cell lines repaired UV-induced damage much more quickly and showed markedly reduced apoptosis compared to their parental counterparts [58]. However, these results have not been confirmed in fresh patient cells.

7.2.3 Cell Cycle Checkpoint in BCR-ABL Cells

Studies have shown that CD133$^+$-glioma stem cells activate cell cycle checkpoint pathway more efficiently compared to CD133$^-$ cells ([59], see Chap. 5). BCR/ABL-positive CML cells can repair DSBs more efficiently than the normal counterparts and eventually survive genotoxic treatment. Elevated levels of drug-induced DSBs are associated with higher activity of checkpoint kinase ATR, and enhanced phosphorylation of histone H2AX. This gamma H2AX eventually starts to disappear in BCR/ABL cells, while continues to increase in parental cells. In addition, expression and ATR-dependent phosphorylation of Chk1 kinase on serine 345 are often more abundant in BCR/ABL cells [55]. Furthermore, BCR/ABL stimulates expression of Nbs1, a member of Mre11-Rad50-Nbs1 complex that plays crucial roles in DNA repair and checkpoint activation. Enhanced ATM-dependent phosphorylation of Nbs1 on serine 343 was observed after damage [60]. A number of other reports have also shown that BCR/ABL-positive cells display enhanced G2-M checkpoint activation in response to various DNA damaging agents including cisplatin, MMC, etoposide and daunorubicin. This enhanced activation of the checkpoint seems to cause resistance to chemotherapies [42, 55, 61–64]. This effect might be due to ATR- and ATM-mediated phosphorylation of p53, leading to its accumulation causing upregulation of p21^{Waf1} and GADD45 [65]. In fact, increased p53 accumulation after DNA damage has been reported in CML primary cells [66]. This effect was associated with ABL kinase-dependent stimulation of ATR/ATM and p53 phosphorylation. Moreover, a checkpoint kinase ATM is shown to phosphorylate c-Abl in response to irradiation [44]. However, in contrast to these observations, one report shows an opposite result. BCR/ABL kinase protein translocates to the nucleus, associates with ATR and disrupts ATR-dependent intra-S-phase checkpoint, leading to a radio-resistant phenotype and prolonged G2-M checkpoint after etoposide treatment [67]. Although the reason for this discrepancy is unknown, the differences in the cells and cell lines used in these studies might be responsible. The latter used an inducible model, and the others used stably expressing BCR/ABL cell lines and/or primary patient cells. Constitutive but not inducible expression of BCR/ABL might better mimic the conditions in established Philadelphia chromosome-positive leukemia cells [60].

As described above, different results have been reported on similar experiments for investigation of DDR pathways in BCR/ABL-expressing cells. It is highly possible that this was caused by differences in cell/cell line system utilized in those studies. The majority of experiments are performed with CD34$^+$-fresh or short-term cultured patient cells from bone marrow, comparing to CD34$^+$ cells from healthy donors. This setting is probably the best for understanding DDR regulation in leukemia and also in leukemia stem/progenitor cells *in vitro*. In some cases, similar results were obtained from fibroblasts or transformed cells with artificial expression of BCR/ABL, indicating that overexpression of the oncoprotein itself effects DDR. Alteration of this response could be caused by the direct effects of BCR/ABL on DNA damage response proteins and/or gene expression, or by induction of DNA damage such as ROS. There have not been comprehensive studies to investigate DDR in true LSCs comparing to progenitor and differentiated cells to date. Such studies will be valuable in unraveling reasons why chemotherapies fail and cause relapses in some cases.

7.3 Cancer Stem Cells in CNS Tumors

The most common and well-characterized CNS tumor is glioblastoma. It still remains controversial, however an enhanced DNA repair capacity and preferential activation of DNA damage checkpoint pathway have been reported in CD133^{+} glioma stem cells ([59], see Chap. 5). Similar results were demonstrated in another CNS tumor, atypical teratoid/rhabdoid tumor (AT/RT) [68]. AT/RT is a rare, aggressive, and highly malignant tumor that commonly occurs in infancy and childhood [69–72]. In the past, the majority of AT/RTs were misclassified as primitive dermal tumors (PNET) and medulloblastoma (MB) at supratentorial sites because of the similarities in radiological and histological features of these tumors [73, 74]. As the word *teratoid* indicates, AT/RTs show multiple-lineage developmental characteristics of malignant teratomas of neuroectodermal, mesodermal, and endodermal lineages [73–75]. Clinical data have indicated that the amount of CD133^{+} cells in AT/RTs correlated positively with degrees of resistance to radiation therapies. Increased phosphorylation of checkpoint proteins, ATM, RAD17 and CHK1 as well as increased expression of BCL-2 in CD133^{+} cells as compared to CD133^{-} cells were observed after radiation. Furthermore, CD133^{+} cells were found to be more resistant to ionizing radiation (IR) in combination with cisplatin-and/or TRAIL-induced apoptosis [68].

Another pediatric CNS tumor medulloblastoma contains CSCs in a perivascular niche. It has been speculated that the CSC population gives rise to recurrence following radiation. A mouse medulloblastoma model showed that the nestin-expressing perivascular stem cells survive radiation, activate PI3K/Akt pathway, undergo p53-dependent cell cycle arrest, and reenter the cell cycle, whereas the proliferating cells in the tumor bulk undergo radiation-induced p53-dependent apoptotic cell death. Activation of Akt signaling via PTEN loss transforms these cells to a non-proliferating extensive nodular morphology [77]. Effects of Akt activation on DNA repair and checkpoint responses were not investigated in the study. However, involvements of Akt in DDR pathways have been demonstrated. Activation of Akt in response to IR and temozolomide depends on ATM and ATR [78, 79]. Activation of the Akt pathway has been linked to chemoresistance in colon and breast cancer cells as well as in CD133^{+} hepatocellular carcinoma [79, 80]. However, another study reported that Akt activation suppresses Chk2-mediated temozolomide-induced G2 arrest in a glioma cell line [81].

7.4 Cancer Stem Cells in Pancreatic and Prostate Cancer

Increased expression of DNA repair genes were found in invasive human pancreatic cancer cells [82–86]. The same trend was observed in other cancers including cervix [87], head and neck [88], brain [89], kidney [90] and bladder [91]. Similar results were obtained from invasive human prostate cancer cells. These cells undergo an epithelial to mesenchymal transition during the process of invasion [92]. In the invasive pancreatic cells, the upregulated genes included BRCA1, FANCI and RAD51. It was demonstrated that the invasive prostate cancer cell population exhibited cancer stem cell-like properties such as high tumorigenicity in mice and elevated expression of stem cell markers [82]. Cells overexpressing RAD51 showed higher rate of survival compared to cells that expressing basal levels of RAD51 after a DSB-inducing drug [93]. Furthermore, overexpression of Rad51 causes dysregulated homologous recombination (HR) and elevated genetic instability [94, 95]. Therefore, Rad51 overexpressing cancer stem cells might acquire survival advantage and accumulate genomic instability leading to progression of tumors. An enhanced level of BRCA1 foci without damage and faster repair after a cytotoxic pyrimidine-analog drug, gemcitabine, treatment were observed [82]. The link between an invasive population of cancer cells and CSCs alls fits within "the cancer stem cell hypothesis" (see Chap. 1). The small population of CSCs has the ability to survive after chemo- and radiotherapies leading to aggressiveness and relapse of tumors. In fact, resistance to gemicitabine was shown to be associated with cancer stem cell-like phenotype, although causes of the resistance were not addressed [96]. One possible reason is enhanced DNA repair and damage response capacity in the population.

7.5 Cancer Stem Cells in Colon Cancer

Consistent with reports in glioma and breast CSCs (see above Chaps. 5 and 6), preferential activation of the checkpoint in $CD133^+$ colon cancer stem cells was recently observed [97]. In this study, enhanced activation of Chk1 was observed after treatment with the intra-crosslinking agent mitomycin C. Inhibition of the ATR but not ATM pathway depleted $CD133^+$ tumorigenic cells *in vitro* and *in vivo*. Caffeine, a non-specific inhibitor of checkpoint-modulating phosphoinositide 3-kinase related (PIK) kinases, increased proliferation and apoptosis of $CD133^+$ colon CSCs. Induction of stalled replication forks by mitomycin C increased the effect of ATR/Chk1 inhibition on the $CD133^+$ population. The Fanconi anemia pathway is required for intra-crosslink DNA repair, and is mediated by the ATR pathway [98, 99]. However, no significant differences in the $CD133^+$ population in FANCC and FANCG deficient cells were observed [97]. ATR has also been shown to be required for normal stem cell maintenance, and furthermore, ATR is an essential gene for embryonic development [100]. However, ATR conditional knockout mice exhibit dramatic reduction of tissue-specific stem and progenitor cells and exhaustion of tissue renewal and homeostatic capacity [101]. Similarly, ATM is required for self-renewal of hematopoietic stem cells, but is not important for proliferation or differentiation of progenitors. ATM knockout mice older than 24 weeks showed progressive bone marrow failure from a defect in HSC function that was associated with elevated ROS [102]. Requirement of ATR but not ATM for tumorigenicity of colon CSCs might be due to its requirement for cell survival. Enhanced activation of the ATM pathway might be observed with different damaging agents such as radiation in the CSCs.

7.6 Cancer Stem Cells in Lung Cancer

Preferential activation of the checkpoint and faster repair were reported in non-small-cell lung cancer (NSCLC) stem cells as compared to differentiated progenies [103]. The authors compared Chk1 activation and gamma H2AX status of NSCLC stem cells and differentiated cells after treatments with various chemotherapeutic agents. Chk1 was activated more efficiently, and much fewer gamma H2AX foci were detected in the CSCs compared to the differentiated counterparts. Furthermore, chemotherapy resistance of NSCLC stem cells was associated with rapid and sustained Chk1 activation regardless of their p53 status. Combination of chemotherapeutic drugs with Chk1 inhibitors prevented DNA repair, suggesting that NSCLC stem cells lose the ability to repair damaged DNA in the presence of Chk1 inhibitors. In contrast, differentiated progenies died after long exposure to chemotherapeutic agents independently of the presence of the Chk1 inhibitors. These data were further confirmed in mouse xenograft models *in vivo*.

$CD133^{+}$ epithelial specific antigen positive ($CD133^{+}$ ESA^{+}) NSCLC stem cells were shown to be highly tumorigenic and were spared by cisplatin treatment [104]. In this study, the DNA damage response in the cancer stem cells was not investigated, but association of the drug resistance with expression of multidrug transporters of the ATP-binding cassette (ABC) superfamily protein ABCG2 was described. In another study, association of radiation resistant cells with presence of ALDH1 but not with other stem cell markers CD133, Sox2 and Oct4 was found [105]. ALDH1 has been discussed as a putative CSC marker for various cancer entities, such as breast, brain, and HNSCC [5, 106–109]. The authors enriched radioresistant cells from a lung cancer cell line, and then investigated whether the radioresitant cells present with CSC characteristics, including enhanced DNA damage response. The radioresitent cells exhibited enhanced DSB repair judged by lower amount of gamma H2AX foci formation after irradiation. Phosphorylation of DNA-PKcs at S2056 was enhanced in the resistant cells compared to the parental cells although the expression level of DNA-PKcs was comparable in both cells [105].

7.7 Future Directions

DDR controls in tissue specific stem cells came into view by recent studies [110–113]. These studies clearly demonstrated the existence of common mechanisms to limit the amount of DNA damage, to restrain them from undergoing massive apoptosis and being exhausted following DNA damage, and to preserve overall tissue function [14]. Quiescent stem cells choose to survive by inhibiting apoptotic pathways and repair damaged DNA by error-prone repair pathways such as NHEJ, leading to accumulation of genomic instability. This mechanism is important to maintain tissue function in the short term, but might meet the long-term consequences such as cancer development, aging, and tissue atrophy. Proliferating stem cells in the cases of umbilical cord blood hematopoietic stem cells, which are still considered to be of fetal origin, and intestinal stem cells choose to undergo massive apoptosis after damage, avoiding accumulation of genomic instabilities.

The cellular origins of CSCs are still under debate. However, speculation exists that CSCs are derived from tissue specific stem or progenitor cells. If that is the case, CSCs might inherit the preferences of DDR pathways of their origin. However, proliferation statuses of CSCs are generally not determined *in vivo,* except in some cases such as leukemic stem cells which have been sown to be quiescent similar to hematopoietic stem cells. It is possible that quiescent stem cells acquire a proliferative status during the process of tumorigenesis. It is an important question to address in order to understand evolution of tumors and also to develop efficient therapies which are toxic to CSCs but not to the normal counterpart.

Unlike tissue specific stem cells, situations in cancer stem cells are much more complicated due to heterogeneous features of cancer cells. DDR and cell proliferation genes are often mutated in cancer cells. The background mutations of tumors might change DDR of cancer stem cells greatly. Furthermore, isolation methods and stem cell markers for solid cancers are not as well defined as LSCs. In most cases, unlike the hematopoietic system, the normal tissue developmental hierarchy has not been identified or characterized. This makes the selection of candidate markers more difficult. These factors might lead to controversial results. Future studies on defined CSCs are required in order to obtain clear results.

Most experiments on DDR of CSCs are performed *in vitro*. However, existence of tissue specific stem cells as well as CSCs require stem cell niches which are often found in perivascular regions. The regions are known to be hypoxic and might induce high levels of ROS, changing the physiology of the cells found in this area. Environments around the stem cell niche might affect DDR of CSCs. Therefore, it is essential to confirm *in vitro* results further *in vivo*.

Although further intensive studies are required, we now recognize enhanced DDR activities in many types of CSCs. It is well accepted that CSCs are a cause of failures and relapses of chemo- and radiotherapies. Chemotherapeutic agents are often DNA damaging agents, and radiation causes DSBs and ROS. The next stage in this field is to compare DDR of malignant (aggressive and invasive) versus benign, primary versus recurrent, and primary versus metastatic or secondary CSCs.

Addressing the questions above will lead us understanding the mechanism of tumor development and revolutionize cancer therapies.

References

1. Ailles LE, Weissman IL (2007) Cancer stem cells in solid tumors. Curr Opin Biotechnol 18(5):460–466
2. Hope KJ, Jin L, Dick JE (2004) Acute myeloid leukemia originates from a hierarchy of leukemic stem cell classes that differ in self-renewal capacity. Nat Immunol 5(7):738–743
3. Al-Hajj M, Wicha MS, Benito-Hernandez A, Morrison SJ, Clarke MF (2003) Prospective identification of tumorigenic breast cancer cells. Proc Natl Acad Sci USA 100(7):3983–3988; PMCID: 153034
4. Singh SK, Hawkins C, Clarke ID, Squire JA, Bayani J, Hide T et al (2004) Identification of human brain tumour initiating cells. Nature 432(7015):396–401
5. Matsui W, Huff CA, Wang Q, Malehorn MT, Barber J, Tanhehco Y et al (2004) Characterization of clonogenic multiple myeloma cells. Blood 103(6):2332–2336
6. Prince ME, Sivanandan R, Kaczorowski A, Wolf GT, Kaplan MJ, Dalerba P et al (2007) Identification of a subpopulation of cells with cancer stem cell properties in head and neck squamous cell carcinoma. Proc Natl Acad Sci USA 104(3):973–8; PMCID: 1783424
7. Li C, Heidt DG, Dalerba P, Burant CF, Zhang L, Adsay V et al (2007) Identification of pancreatic cancer stem cells. Cancer Res 67(3):1030–1037
8. Dalerba P, Dylla SJ, Park IK, Liu R, Wang X, Cho RW et al (2007) Phenotypic characterization of human colorectal cancer stem cells. Proc Natl Acad Sci USA 104(24):10158–10163; PMCID: 1891215
9. O'Brien CA, Pollett A, Gallinger S, Dick JE (2007) A human colon cancer cell capable of initiating tumour growth in immunodeficient mice. Nature 445(7123):106–110
10. Reynolds BA, Weiss S (1996) Clonal and population analyses demonstrate that an EGF-responsive mammalian embryonic CNS precursor is a stem cell. Dev Biol 175(1):1–13
11. Fang D, Nguyen TK, Leishear K, Finko R, Kulp AN, Hotz S et al (2005) A tumorigenic subpopulation with stem cell properties in melanomas. Cancer Res 65(20):9328–9337
12. Collins AT, Berry PA, Hyde C, Stower MJ, Maitland NJ (2005) Prospective identification of tumorigenic prostate cancer stem cells. Cancer Res 65(23):10946–10951
13. Singh SK, Clarke ID, Terasaki M, Bonn VE, Hawkins C, Squire J et al (2003) Identification of a cancer stem cell in human brain tumors. Cancer Res 63(18):5821–5828
14. Blanpain C, Mohrin M, Sotiropoulou PA, Passegue E (2011) DNA-damage response in tissue-specific and cancer stem cells. Cell Stem Cell 8(1):16–29
15. Bonnet D, Dick JE (1997) Human acute myeloid leukemia is organized as a hierarchy that originates from a primitive hematopoietic cell. Nat Med 3(7):730–737
16. Passegue E, Weisman IL (2005) Leukemic stem cells: where do they come from? Stem Cell Rev 1(3):181–188
17. Passegue E (2005) Hematopoietic stem cells, leukemic stem cells and chronic myelogenous leukemia. Cell Cycle 4(2):266–268
18. Jamieson CH, Ailles LE, Dylla SJ, Muijtjens M, Jones C, Zehnder JL et al (2004) Granulocyte-macrophage progenitors as candidate leukemic stem cells in blast-crisis CML. N Engl J Med 351(7):657–667
19. Guzman ML, Jordan CT (2004) Considerations for targeting malignant stem cells in leukemia. Cancer Control 11(2):97–104
20. Guan Y, Gerhard B, Hogge DE (2003) Detection, isolation, and stimulation of quiescent primitive leukemic progenitor cells from patients with acute myeloid leukemia (AML). Blood 101(8):3142–3149
21. Holyoake T, Jiang X, Eaves C, Eaves A (1999) Isolation of a highly quiescent subpopulation of primitive leukemic cells in chronic myeloid leukemia. Blood 94(6):2056–2064
22. Guzman ML, Jordan CT (2009) Lessons learned from the study of JunB: new insights for normal and leukemia stem cell biology. Cancer Cell 15(4):252–254
23. Elrick LJ, Jorgensen HG, Mountford JC, Holyoake TL (2005) Punish the parent not the progeny. Blood 105(5):1862–1866

24. Wong S, Witte ON (2004) The BCR-ABL story: bench to bedside and back. Annu Rev Immunol 22:247–306
25. Peggs K, Mackinnon S (2003) Imatinib mesylate—the new gold standard for treatment of chronic myeloid leukemia. N Engl J Med 348(11):1048–1050
26. Calabretta B, Perrotti D (2004) The biology of CML blast crisis. Blood 103(11):4010–4022
27. Melo JV, Barnes DJ (2007) Chronic myeloid leukaemia as a model of disease evolution in human cancer. Nature Rev Cancer 7(6):441–453
28. Yong AS, Melo JV (2009) The impact of gene profiling in chronic myeloid leukaemia. Best Pract Res Clin Haematol 22(2):181–190
29. Radich JP, Dai H, Mao M, Oehler V, Schelter J, Druker B et al (2006) Gene expression changes associated with progression and response in chronic myeloid leukemia. Proc Natl Acad Sci USA 103(8):2794–2799; PMCID: 1413797
30. Oehler VG, Guthrie KA, Cummings CL, Sabo K, Wood BL, Gooley T et al (2009) The preferentially expressed antigen in melanoma (PRAME) inhibits myeloid differentiation in normal hematopoietic and leukemic progenitor cells. Blood 114(15):3299–3308; PMCID: 2759652
31. Perrotti D, Jamieson C, Goldman J, Skorski T (2010) Chronic myeloid leukemia: mechanisms of blastic transformation. J Clin Invest 120(7):2254–2264; PMCID: 2898591
32. Schultheis B, Szydlo R, Mahon FX, Apperley JF, Melo JV (2005) Analysis of total phosphotyrosine levels in $CD34^+$ cells from CML patients to predict the response to imatinib mesylate treatment. Blood 105(12):4893–4894
33. Barnes DJ, Palaiologou D, Panousopoulou E, Schultheis B, Yong AS, Wong A et al (2005) Bcr-Abl expression levels determine the rate of development of resistance to imatinib mesylate in chronic myeloid leukemia. Cancer Res 65(19):8912–8919
34. Perrotti D, Cesi V, Trotta R, Guerzoni C, Santilli G, Campbell K et al (2002) BCR-ABL suppresses C/EBPalpha expression through inhibitory action of hnRNP E2. Nat Genet 30(1):48–58
35. Burke BA, Carroll M (2010) BCR-ABL: a multi-faceted promoter of DNA mutation in chronic myelogeneous leukemia. Leukemia 24(6):1105–1112
36. Cramer K, Nieborowska-Skorska M, Koptyra M, Slupianek A, Penserga ET, Eaves CJ et al (2008) BCR/ABL and other kinases from chronic myeloproliferative disorders stimulate single-strand annealing, an unfaithful DNA double-strand break repair. Cancer Res 68(17):6884–6888; PMCID: 2531069
37. Koptyra M, Falinski R, Nowicki MO, Stoklosa T, Majsterek I, Nieborowska-Skorska M et al (2006) BCR/ABL kinase induces self-mutagenesis via reactive oxygen species to encode imatinib resistance. Blood 108(1):319–327; PMCID: 1895841
38. Nowicki MO, Falinski R, Koptyra M, Slupianek A, Stoklosa T, Gloc E et al (2004) BCR/ABL oncogenic kinase promotes unfaithful repair of the reactive oxygen species-dependent DNA double-strand breaks. Blood 104(12):3746–3753
39. Bernstein C, Bernstein H, Payne CM, Garewal H (2002) DNA repair/pro-apoptotic dual-role proteins in five major DNA repair pathways: fail-safe protection against carcinogenesis. Mutat Res 511(2):145–178
40. Bernstein R (1988) Cytogenetics of chronic myelogenous leukemia. Semin Hematol 25(1):20–34
41. Slupianek A, Nowicki MO, Koptyra M, Skorski T (2006) BCR/ABL modifies the kinetics and fidelity of DNA double-strand breaks repair in hematopoietic cells. DNA Repair 5(2):243–250; PMCID: 2856314
42. Slupianek A, Hoser G, Majsterek I, Bronisz A, Malecki M, Blasiak J et al (2002) Fusion tyrosine kinases induce drug resistance by stimulation of homology-dependent recombination repair, prolongation of G(2)/M phase, and protection from apoptosis. Mol Cell Biol 22(12):4189–4201; PMCID: 133854
43. Deutsch E, Jarrousse S, Buet D, Dugray A, Bonnet ML, Vozenin-Brotons MC et al (2003) Down-regulation of BRCA1 in BCR-ABL-expressing hematopoietic cells. Blood 101(11):4583–4588

44. Chen G, Yuan SS, Liu W, Xu Y, Trujillo K, Song B et al (1999) Radiation-induced assembly of Rad51 and Rad52 recombination complex requires ATM and c-Abl. J Biol Chem 274(18):12748–12752
45. Pastwa E, Poplawski T, Czechowska A, Malinowski M, Blasiak J (2005) Non-homologous DNA end joining repair in normal and leukemic cells depends on the substrate ends. Z Naturforsch C 60(5–6):493–500
46. Deutsch E, Dugray A, AbdulKarim B, Marangoni E, Maggiorella L, Vaganay S et al (2001) BCR-ABL down-regulates the DNA repair protein DNA-PKcs. Blood 97(7):2084–2090
47. Skorski T (2002) BCR/ABL regulates response to DNA damage: the role in resistance to genotoxic treatment and in genomic instability. Oncogene 21(56):8591–8604
48. Shah NP, Sawyers CL (2003) Mechanisms of resistance to STI571 in Philadelphia chromosome-associated leukemias. Oncogene 22(47):7389–7395
49. Flamant S, Turhan AG (2005) Occurrence of de novo ABL kinase domain mutations in primary bone marrow cells after BCR-ABL gene transfer and Imatinib mesylate selection. Leukemia 19(7):1265–1267
50. von Bubnoff N, Barwisch S, Speicher MR, Peschel C, Duyster J (2005) A cell-based screening strategy that predicts mutations in oncogenic tyrosine kinases: implications for clinical resistance in targeted cancer treatment. Cell Cycle 4(3):400–406
51. Stoklosa T, Poplawski T, Koptyra M, Nieborowska-Skorska M, Basak G, Slupianek A et al (2008) BCR/ABL inhibits mismatch repair to protect from apoptosis and induce point mutations. Cancer Res 68(8):2576–2580
52. Wada C, Shionoya S, Fujino Y, Tokuhiro H, Akahoshi T, Uchida T et al (1994) Genomic instability of microsatellite repeats and its association with the evolution of chronic myelogenous leukemia. Blood 83(12):3449–3456
53. Maru Y, Kobayashi T, Tanaka K, Shibuya M (1999) BCR binds to the xeroderma pigmentosum group B protein. Biochem Biophys Res Commun 260(2):309–312
54. Takeda N, Shibuya M, Maru Y (1999) The BCR-ABL oncoprotein potentially interacts with the xeroderma pigmentosum group B protein. Proc Natl Acad Sci USA 96(1):203–255. PMCID: 15117
55. Nieborowska-Skorska M, Stoklosa T, Datta M, Czechowska A, Rink L, Slupianek A et al (2006) ATR-Chk1 axis protects BCR/ABL leukemia cells from the lethal effect of DNA double-strand breaks. Cell Cycle 5(9):994–1000
56. Maru Y, Bergmann E, Coin F, Egly JM, Shibuya M (2001) TFIIH functions are altered by the P210BCR-ABL oncoprotein produced on the Philadelphia chromosome. Mutation Research 483(1–2):83–88
57. Canitrot Y, Falinski R, Louat T, Laurent G, Cazaux C, Hoffmann JS et al (2003) p210 BCR/ABL kinase regulates nucleotide excision repair (NER) and resistance to UV radiation. Blood 102(7):2632–2637
58. Laurent E, Mitchell DL, Estrov Z, Lowery M, Tucker SL, Talpaz M et al (2003) Impact of p210(Bcr-Abl) on ultraviolet C wavelength-induced DNA damage and repair. Clin Cancer Res 9(10 Pt 1):3722–3730
59. Bao S, Wu Q, McLendon RE, Hao Y, Shi Q, Hjelmeland AB et al (2006) Glioma stem cells promote radioresistance by preferential activation of the DNA damage response. Nature 444(7120):756–760
60. Rink L, Slupianek A, Stoklosa T, Nieborowska-Skorska M, Urbanska K, Seferynska I et al (2007) Enhanced phosphorylation of Nbs1, a member of DNA repair/checkpoint complex Mre11-RAD50-Nbs1, can be targeted to increase the efficacy of imatinib mesylate against BCR/ABL-positive leukemia cells. Blood 110(2):651–660; PMCID: 1924483
61. Bedi A, Barber JP, Bedi GC, el-Deiry WS, Sidransky D, Vala MS et al (1995) BCR-ABL-mediated inhibition of apoptosis with delay of G2/M transition after DNA damage: a mechanism of resistance to multiple anticancer agents. Blood 86(3):1148–1158
62. Nishii K, Kabarowski JH, Gibbons DL, Griffiths SD, Titley I, Wiedemann LM et al (1996) ts BCR-ABL kinase activation confers increased resistance to genotoxic damage via cell cycle block. Oncogene 13(10):2225–2234

63. Stiewe T, Parssanedjad K, Esche H, Opalka B, Putzer BM (2000). E1A overcomes the apoptosis block in BCR-ABL^{+} leukemia cells and renders cells susceptible to induction of apoptosis by chemotherapeutic agents. Cancer Res 60(14):3957–3964
64. Higginbottom K, Cummings M, Newland AC, Allen PD (2002) Etoposide-mediated deregulation of the G2M checkpoint in myeloid leukaemic cell lines results in loss of cell survival. Br J Haematol 119(4):956–964
65. Stoklosa T, Slupianek A, Datta M, Nieborowska-Skorska M, Nowicki MO, Koptyra M et al (2004) BCR/ABL recruits p53 tumor suppressor protein to induce drug resistance. Cell Cycle 3(11):1463–1472
66. Goldberg Z, Levav Y, Krichevsky S, Fibach E, Haupt Y (2004) Treatment of chronic myeloid leukemia cells with imatinib (STI571) impairs p53 accumulation in response to DNA damage. Cell Cycle 3(9):1188–1195
67. Dierov J, Dierova R, Carroll M (2004) BCR/ABL translocates to the nucleus and disrupts an ATR-dependent intra-S phase checkpoint. Cancer Cell 5(3):275–285
68. Chiou SH, Kao CL, Chen YW, Chien CS, Hung SC, Lo JF et al (2008) Identification of CD133-positive radioresistant cells in atypical teratoid/rhabdoid tumor. PLoS One 3(5):e2090; PMCID: 2396792
69. Sobel EL, Gilles FH, Leviton A, Tavare CJ, Hedley-Whyte ET, Rorke LB et al (1996) Survival of children with infratentorial neuroglial tumors. The Childhood Brain Tumor Consortium. Neurosurgery 39(1):45–54; discussion 54–6
70. Burger PC, Yu IT, Tihan T, Friedman HS, Strother DR, Kepner JL et al (1998) Atypical teratoid/rhabdoid tumor of the central nervous system: a highly malignant tumor of infancy and childhood frequently mistaken for medulloblastoma: a Pediatric Oncology Group study. Am J Surg Pathol 22(9):1083–1092
71. Tekautz TM, Fuller CE, Blaney S, Fouladi M, Broniscer A, Merchant TE et al (2005) Atypical teratoid/rhabdoid tumors (ATRT): improved survival in children 3 years of age and older with radiation therapy and high-dose alkylator-based chemotherapy. J Clin Oncol 23(7):1491–1499
72. Wong TT, Ho DM, Chang KP, Yen SH, Guo WY, Chang FC et al (2005) Primary pediatric brain tumors: statistics of Taipei VGH, Taiwan (1975–2004). Cancer 104(10):2156–2167
73. Parwani AV, Stelow EB, Pambuccian SE, Burger PC, Ali SZ (2005) Atypical teratoid/rhabdoid tumor of the brain: cytopathologic characteristics and differential diagnosis. Cancer 105(2):65–70
74. Cheng YC, Lirng JF, Chang FC, Guo WY, Teng MM, Chang CY et al (2005) Neuroradiological findings in atypical teratoid/rhabdoid tumor of the central nervous system. Acta Radiol 46(1):89–96
75. Bergmann M, Spaar HJ, Ebhard G, Masini T, Edel G, Gullotta F et al (1997) Primary malignant rhabdoid tumours of the central nervous system: an immunohistochemical and ultrastructural study. Acta Neurochir (Wien) 139(10):961–968; discussion 8–9
76. Ho DM, Hsu CY, Wong TT, Ting LT, Chiang H (2000) Atypical teratoid/rhabdoid tumor of the central nervous system: a comparative study with primitive neuroectodermal tumor/medulloblastoma. Acta Neuropathol 99(5):482–488
77. Hambardzumyan D, Becher OJ, Rosenblum MK, Pandolfi PP, Manova-Todorova K, Holland EC (2008) PI3K pathway regulates survival of cancer stem cells residing in the perivascular niche following radiation in medulloblastoma in vivo. Gene Dev 22(4):436–448; PMCID: 2238666
78. Viniegra JG, Martinez N, Modirassari P, Losa JH, Parada Cobo C, Lobo VJ et al (2005) Full activation of PKB/Akt in response to insulin or ionizing radiation is mediated through ATM. J Biol Chem 280(6):4029–4036
79. Caporali S, Levati L, Starace G, Ragone G, Bonmassar E, Alvino E et al (2008) AKT is activated in an ataxia-telangiectasia and Rad3-related-dependent manner in response to temozolomide and confers protection against drug-induced cell growth inhibition. Mol Pharmacol 74(1):173–183
80. Ma S, Lee TK, Zheng BJ, Chan KW, Guan XY (2008) CD133^{+} HCC cancer stem cells confer chemoresistance by preferential expression of the Akt/PKB survival pathway. Oncogene 27(12):1749–1758

81. Hirose Y, Katayama M, Mirzoeva OK, Berger MS, Pieper RO (2005) Akt activation suppresses Chk2-mediated, methylating agent-induced G2 arrest and protects from temozolomide-induced mitotic catastrophe and cellular senescence. Cancer Res 65(11):4861–4869
82. Mathews LA, Cabarcas SM, Hurt EM, Zhang X, Jaffee EM, Farrar WL (2011) Increased expression of DNA repair genes in invasive human pancreatic cancer cells. Pancreas 40(5):730–739; PMCID: 3116046
83. Yu J, Rhodes DR, Tomlins SA, Cao X, Chen G, Mehra R et al (2007) A polycomb repression signature in metastatic prostate cancer predicts cancer outcome. Cancer Res 67(22):10657–10663
84. Lapointe J, Li C, Higgins JP, van de Rijn M, Bair E, Montgomery K et al (2004) Gene expression profiling identifies clinically relevant subtypes of prostate cancer. Proc Natl Acad Sci USA 101(3):811–816; PMCID: 321763
85. LaTulippe E, Satagopan J, Smith A, Scher H, Scardino P, Reuter V et al (2002) Comprehensive gene expression analysis of prostate cancer reveals distinct transcriptional programs associated with metastatic disease. Cancer Res 62(15):4499–4506
86. Varambally S, Yu J, Laxman B, Rhodes DR, Mehra R, Tomlins SA et al (2005) Integrative genomic and proteomic analysis of prostate cancer reveals signatures of metastatic progression. Cancer Cell 8(5):393–406
87. Pyeon D, Newton MA, Lambert PF, den Boon JA, Sengupta S, Marsit CJ et al (2007) Fundamental differences in cell cycle deregulation in human papillomavirus-positive and human papillomavirus-negative head/neck and cervical cancers. Cancer Res 67(10):4605–4619; PMCID: 2858285
88. Schlingemann J, Habtemichael N, Ittrich C, Toedt G, Kramer H, Hambek M et al (2005) Patient-based cross-platform comparison of oligonucleotide microarray expression profiles. Lab Invest 85(8):1024–1039
89. Albino D, Scaruffi P, Moretti S, Coco S, Truini M, Di Cristofano C et al (2008) Identification of low intratumoral gene expression heterogeneity in neuroblastic tumors by genome-wide expression analysis and game theory. Cancer 113(6):1412–1422
90. Yusenko MV, Kuiper RP, Boethe T, Ljungberg B, van Kessel AG, Kovacs G (2009) High-resolution DNA copy number and gene expression analyses distinguish chromophobe renal cell carcinomas and renal oncocytomas. BMC Cancer 9:152; PMCID: 2686725
91. Sanchez-Carbayo M, Socci ND, Lozano J, Saint F, Cordon-Cardo C (2006) Defining molecular profiles of poor outcome in patients with invasive bladder cancer using oligonucleotide microarrays. J Clin Oncol 24(5):778–789
92. Klarmann GJ, Hurt EM, Mathews LA, Zhang X, Duhagon MA, Mistree T et al (2009) Invasive prostate cancer cells are tumor initiating cells that have a stem cell-like genomic signature. Clin Exp Metastasis 26(5):433–446; PMCID: 2782741
93. Maacke H, Jost K, Opitz S, Miska S, Yuan Y, Hasselbach L et al (2000) DNA repair and recombination factor Rad51 is over-expressed in human pancreatic adenocarcinoma. Oncogene 19(23):2791–2795
94. Shammas MA, Shmookler Reis RJ, Koley H, Batchu RB, Li C, Munshi NC (2009) Dysfunctional homologous recombination mediates genomic instability and progression in myeloma. Blood 113(10):2290–2297; PMCID: 2652372
95. Pal J, Bertheau R, Buon L, Qazi A, Batchu RB, Bandyopadhyay S et al (2011) Genomic evolution in Barrett's adenocarcinoma cells: critical roles of elevated hsRAD51, homologous recombination and Alu sequences in the genome. Oncogene 30(33):3585–3598
96. Hu G, Li F, Ouyang K, Xie F, Tang X, Wang K et al (2011) Intrinsic gemcitabine resistance in a novel pancreatic cancer cell line. Int J Oncol 40:798–806
97. Gallmeier E, Hermann PC, Mueller MT, Machado JG, Ziesch A, De Toni EN et al (2011) Inhibition of ataxia telangiectasia- and Rad3-related function abrogates the in vitro and in vivo tumorigenicity of human colon cancer cells through depletion of the CD133(+) tumor-initiating cell fraction. Stem Cells 29(3):418–429
98. Grompe M, D'Andrea A (2001) Fanconi anemia and DNA repair. Hum Mol Genet 10(20):2253–2259

99. Andreassen PR, D'Andrea AD, Taniguchi T (2004) ATR couples FANCD2 monoubiquitination to the DNA-damage response. Gene Dev 18(16):1958–1963; PMCID: 514175
100. Brown EJ, Baltimore D (2000) ATR disruption leads to chromosomal fragmentation and early embryonic lethality. Gene Dev 14(4):397–402; PMCID: 316378
101. Ruzankina Y, Pinzon-Guzman C, Asare A, Ong T, Pontano L, Cotsarelis G et al (2007) Deletion of the developmentally essential gene ATR in adult mice leads to age-related phenotypes and stem cell loss. Cell Stem Cell 1(1):113–126; PMCID: 2920603
102. Ito K, Hirao A, Arai F, Matsuoka S, Takubo K, Hamaguchi I et al (2004) Regulation of oxidative stress by ATM is required for self-renewal of haematopoietic stem cells. Nature 431(7011):997–1002
103. Bartucci M, Svensson S, Romania P, Dattilo R, Patrizii M, Signore M et al (2011) Therapeutic targeting of Chk1 in NSCLC stem cells during chemotherapy. Cell Death Differ 19:768–778
104. Bertolini G, Roz L, Perego P, Tortoreto M, Fontanella E, Gatti L et al (2009) Highly tumorigenic lung cancer $CD133^{+}$ cells display stem-like features and are spared by cisplatin treatment. Proc Natl Acad Sci USA 106(38):16281–16286; PMCID: 2741477
105. Mihatsch J, Toulany M, Bareiss PM, Grimm S, Lengerke C, Kehlbach R et al (2011) Selection of radioresistant tumor cells and presence of ALDH1 activity in vitro. Radiother Oncol 99(3):300–306
106. Chen YC, Chen YW, Hsu HS, Tseng LM, Huang PI, Lu KH et al (2009) Aldehyde dehydrogenase 1 is a putative marker for cancer stem cells in head and neck squamous cancer. Biochem Bioph Res Co 385(3):307–313
107. Ginestier C, Hur MH, Charafe-Jauffret E, Monville F, Dutcher J, Brown M et al (2007) ALDH1 is a marker of normal and malignant human mammary stem cells and a predictor of poor clinical outcome. Cell Stem Cell 1(5):555–67; PMCID: 2423808
108. Balicki D (2007) Moving Forward in Human Mammary Stem Cell Biology and Breast Cancer Prognostication Using ALDH1. Cell Stem Cell 1(5):485–487
109. Rasper M, Schafer A, Piontek G, Teufel J, Brockhoff G, Ringel F et al (2010) Aldehyde dehydrogenase 1 positive glioblastoma cells show brain tumor stem cell capacity. Neuro Oncol 12(10):1024–1033; PMCID: 3018920
110. Milyavsky M, Gan OI, Trottier M, Komosa M, Tabach O, Notta F et al (2010) A distinctive DNA damage response in human hematopoietic stem cells reveals an apoptosis-independent role for p53 in self-renewal. Cell Stem Cell 7(2):186–197
111. Mohrin M, Bourke E, Alexander D, Warr MR, Barry-Holson K, Le Beau MM et al (2010) Hematopoietic stem cell quiescence promotes error-prone DNA repair and mutagenesis. Cell Stem Cell 7(2):174–185; PMCID: 2924905
112. Blanpain C (2010) Stem cells: skin regeneration and repair. Nature 464(7289):686–687
113. Sotiropoulou PA, Candi A, Mascre G, De Clercq S, Youssef KK, Lapouge G et al (2010) Bcl-2 and accelerated DNA repair mediates resistance of hair follicle bulge stem cells to DNA-damage-induced cell death. Nat cell Biol 12(6):572–582

Chapter 8
Pancreatic Cancer Stem Cells in Tumor Progression, Metastasis, Epithelial-Mesenchymal Transition and DNA Repair

Nagaraj S. Nagathihalli and Erika T. Brown

Abstract Pancreatic cancer is an aggressive solid malignancy with poor response to therapy and the subsequent dismal survival rate has remained a hallmark of this disease. There is evidence to indicate that pancreatic cancer is initiated and propagated by cancer stem cell (CSC)s. The CSC population is defined by its tumor initiating capacity and has been shown to be invasive or metastatic. Loss of genome stability is a hallmark of cancer with DNA repair enzymes aiding in maintenance of stability. The potential to assess the risk of cancer development lies in careful determination of one's capacity in nurturing genome stability. DNA repair genes are over expressed in CSCs and both pancreatic CSCs and invasive cells in turn provide greater DNA damage response and repair mechanisms. Pancreatic tumor-initiating cells as well as invasive cells have a large number of genes related to DNA repair. RAD51, the key player in the recombinational repair of damaged DNA might act as a critical mediator of efficient DNA repair mechanisms of CSCs. We update here the current research results regarding CSCs in pancreatic cancer progression, metastasis and discuss the DNA repair mechanism in pancreatic CSCs.

Keywords Cancer stem cells · DNA repair · EMT · Metastasis · Pancreatic cancer · RAD51

Abbreviations

ALDH	Aldehyde dehydrogenase
BER	Base excision repair
CSCs	Cancer stem cells
CXCR4	CXC chemokine receptor 4
DSB	Double-strand break
EMT	Epithelial to mesenchymal transition
EpCAM	Epithelial cell adhesion molecule

N. S. Nagathihalli (✉)
Department of Surgery, Vanderbilt University School of Medicine,
1161, 21st Ave S., D2300 MCN, Nashville, TN 37232, USA
e-mail: nagaraj.nagathihalli@vanderbilt.edu

E. T. Brown
Department of Pathology & Laboratory Medicine,
Medical University of South Carolina, Charleston, SC 29425, USA

L. A. Mathews et al. (eds.), *DNA Repair of Cancer Stem Cells*,
DOI 10.1007/978-94-007-4590-2_8, © Springer Science+Business Media Dordrecht 2013

ESA	Epithelial specific antigen
HMG CoA	3-hydroxy-3-methylglutaryl coenzyme A reductase
HR	Homologous recombination
MMR	Mismatch repair
NER	Nucleotide excision repair
NHEJ	Non-homologous end-joining
PARP	Poly ADP ribose polymerase
PDAC	Pancreatic ductal adenocarcinoma
SCs	Stem cells
SDF-1	Stromal derived factor-1
SSB	Single-strand break
TGF-β	Transforming growth factor beta
ZEB	Zinc-finger transcription factor

8.1 Introduction

Pancreatic cancer is one of the major causes of cancer death worldwide with an estimated 227,000 deaths each year [1]. An estimated 43,030 new cases will be diagnosed and 37,660 deaths are expected in the United States in 2011. Clinical outcome has not improved substantially over the past 25 years, with an overall 5-year survival rate remaining dismally poor at 5 %. Over the past three decades there has been considerable progress towards understanding the biology of pancreatic cancer and more recently a focus on biomarkers to enable targeted therapies. In spite of these advances the overall survival figures for pancreatic cancer remain bleak and unaltered.

Cancer stem cells (CSCs) are a reservoir of self-sustaining cells that constitute a small subset of cancer cells. The self-renewing properties of CSCs are to fuel to the maintenance of tumor amplification or mass [2, 3]. This property confers CSCs with the potential to initiate secondary tumors bearing similar functions of the parental cells. CSCs have been identified and isolated in nearly all human cancers including breast, lung, prostate, colon, brain, head and neck, liver and pancreas [4–8]. When CSCs are implanted into immunodeficient mice they have the ability to self-renew, differentiate, and regenerate to phenotypic cell types which composed the original tumor [4]. However, the origin of CSCs is still being debated. Potentially, CSCs may originate from genomic instability. The non-tumorigenic stem cells (SCs) spontaneously transform into CSCs through genomic instability or tissue resident SCs that acquire a malignant phenotype by accumulation of mutations. Up to now, pancreatic CSC markers remain controversial.

DNA repair is essential in maintaining genomic integrity and stability, thereby preventing the origination/development of CSCs. Once damage has occurred, multiple pathways are necessary to restore the DNA structure and sequence. The response to DNA damage is often specific to the type of damage incurred and is highly regulated. Exogenous and endogenous factors can cause stranded breaks resulting in fallacious DNA repair and contributing to genomic alterations [9–12]. There are five main repair pathways to maintain the genomic stability: the nucleotide excision repair (NER); base excision repair (BER); mismatch repair (MMR); and the two double-strand break (DSB) repair pathways, non-homologous end-joining (NHEJ), which require no significant homology and homologous recombination (HR) that uses intact sequences on the sister chromatid or the homologous chromosome as a template to repair the broken DNA [13–16]. In DNA repair pathways, a variety of proteins or genes such as ATM, BRCA1, BRCA2, p53, and a number of RAD proteins have been identified as cancer susceptible genes and have been shown to be involved causing mutations [17, 18]. There is an increase in the expression of these DNA repair-related genes in CSCs which renders the cells extremely stable to genetic stability and increases their ability to survive and function as tumor initiators. In tumor metastasis, a higher expression of DNA repair genes confers higher efficiency in repairing damage caused by cytotoxic treatment regimes [19]. DNA repair mechanisms are more efficient in CSCs. There is evidence to support the idea that tumor initiating CSCs are responsible for tumor metastasis [20]. These investigations prompted us to further explore in the published reports for increased DNA repair in pancreatic CSCs populations.

8.2 Pancreatic CSCs in Tumor Progression and Metastasis

Pancreatic cancer has a poor prognosis, and treatment strategies which are based on preclinical investigation have not succeeded in significantly extending patient survival. At the time of diagnosis, only 20 % of the patients suffering from pancreatic cancer present with localized disease amenable to surgery. Forty percent of the patients present with locally advanced disease, and another 40 % already suffer from distant metastases [21]. The pancreatic tumor model demonstrated that mesenchymal SCs play an important role in pancreatic cancer progression [21] (Fig. 8.1). Pancreatic CSCs have been defined by different combinations of markers. These CSCs represent less than 1 % of all pancreatic cancer cells. Li et al., performed the first initial experiments with pancreatic CSCs by isolating EpCAM$^+$CD44$^+$CD24$^+$ cancer cells with high tumorigenic potential using a xenograft model of immunocompromised mice for primary human pancreatic adenocarcinoma [22]. In a second study, Hermann et al., evidenced that CD133 expression as a marker to isolate pancreatic cancer cells with a significantly higher tumorigenic potential [23]. Furthermore, Rasheed et al., recently identified pancreatic cancer cells with ALDH$^+$ stem cell like tumorigenicity, high tumorigenic potential and characteristics of epithelial-mesenchymal transition (EMT) [24]. Interestingly, many isolated and studied CSCs expressed EpCAM including CSCs from breast, colon, prostate and pancreas tumors [22, 25]. This is related to its major role in the cell cycle, proliferation and metabolic function by inducing proto-oncogene *c-myc* and the cell cycle regulating genes cyclin A and E [26]. Recently, the Simeone group identified a CD44$^+$CD24$^+$ESA$^+$ subpopulation as putative pancreatic CSCs in an orthotopic model [22] and the CD44$^+$CD24$^+$ ESA$^+$ pancreatic cancer cells showed the SC properties of self-renewal, the ability to

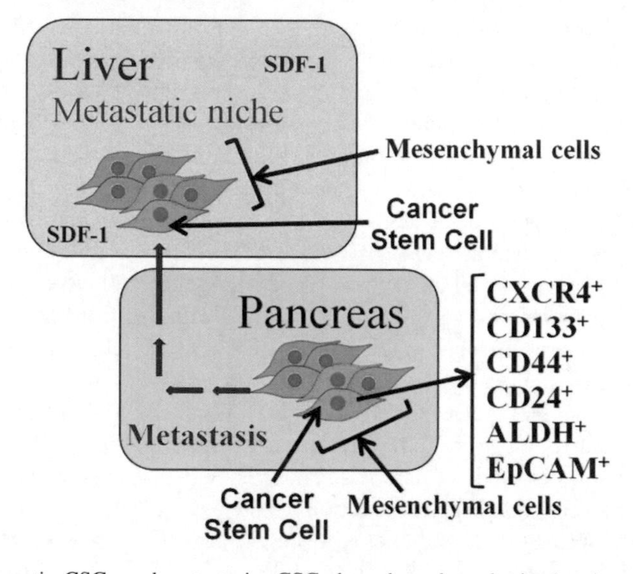

Fig. 8.1 Pancreatic CSCs and metastasis. CSCs have been hypothesized to be responsible for metastatic disease. The most common sites of metastasis of pancreatic adenocarcinoma is liver. Several studies indicate that mesenchymal stem cells (MSCs) support tumor metastasis not only at primary site but also at metastatic sites. These MSCs may lead to a higher incidence of liver metastases and the MSCs over expressing SDF-1 may promote liver metastases. CSC biology in pancreatic cancer has started with identification of multiple CSC phenotypes, such as $CD44^+$, $CD24^+$, $CD133^+$, epithelial cell adhesion molecule ($EpCAM^+$, also known as epithelial specific antigen or ESA^+) and aldehyde dehydrogenase ($ALDH^+$) cells. Pancreatic CSCs expressing CXCR4, a chemokine receptor for the ligand stromal derived factor-1 (SDF-1), were found to be more invasive and mediated metastasis. $CD133^+$ is a marker of CSCs with a higher tumorigenic potential and $CD133^+CXCR4^+$ cancer cells have a higher metastatic potential. CSCs isolated displaying $EpCAM^+CD44^+CD24^+$ have higher tumorigenic potential. $ALDH^+$ stem cells have tumorigenic potential and also display characteristics of EMT

produce differentiated progeny, and increased expression of the signaling molecule sonic hedgehog. This signaling pathway has an early and critical role in the genesis of pancreatic cancer [27].

The relationship between $CD133^+$ pancreatic CSCs, CXCR4 expression, and metastasis has been recently explored [23]. Pancreatic adenocarcinoma also contains 1–3 % of $CD133^+$ cancer cells, some of which also show high expression of CXC chemokine receptor 4 (CXCR4) [8]. In pancreatic cancer, a subset of CSCs expressing CXCR4, a CR for the ligand stromal derived factor-1 (SDF-1), a well studied mediator of cell migration [28–31], were found to be more invasive and mediated metastasis [23] (Fig. 8.1). These CXCR4 and SDF-1 are necessary in the maintenance of pancreatic duct survival, proliferation, and migration during pancreatic organogenesis and regeneration [24]. $CD133^+$ pancreatic CSC were found to be co-expressed with CXCR4 particularly at the invading front of pancreatic tumors. In this study, only $CD133^+CXCR4^+$ cells were able to metastasize, although both $CD133^+CXCR4^-$ and $CD133^+CXCR4^+$ cells were able to form primary tumors equally [23]. This result suggests the existence of stationary and migratory forms of CSCs, which are two distinct phenotypes. This result was extended to clinical specimens where Hermann et al., found significantly higher numbers of $CD133^{+}CXCR4^{+}$ migrating CSCs in patients with lymph node metastasis ($pN1^{+}$) [23]. This demonstrated a close clinical correlation between migrating CSCs and advanced disease. Additionally, $ALDH^{+}$ pancreatic CSCs were found to be invasive *in vitro* and were identified in metastatic lesions when compared with matched primary tumors in another study [24]. Immunohistochemistry has been used to examine the correlation of patient's outcome with the frequency of CSCs in pancreatic cancer and this analysis of ALDH was present in patients with pancreatic cancer and with worse clinical outcomes [24].

The two main properties of CSCs are the capacity for self-renewal as well as the ability to differentiate into a heterogeneous cancer cell population. This process must be recapitulated in distant metastasis as well. Importantly, the same metastatic potential is not equivalent among all cells within a tumor or even among the CSC population. Pancreatic CSCs may have mesenchymal SCs within the tumor and may produce increased levels of the chemotactic cytokines leading to a higher incidence of liver metastasis [21]. It has been proposed that CSCs represent the only cell population capable of spreading and resulting in a metastasis. It seems likely that CSCs, which are progenitors of tumor cells, would be responsible for metastatic spread since metastases can be formed from implantation of a single tumor cell [32]. Chemoresistant cells from pancreatic cancer patients are more tumorigenic and have greater metastatic potential than chemosensitive cancer cells [33, 34]. Earlier reports in pancreatic CSCs have proved that these cells are more tumorigenic and highly resistant to drug therapy [23]. Moreover, Hermann et al., found a distinct subpopulation of migratory CSCs in the invasive front of pancreatic tumors and inhibiting these cell populations almost completely inhibited metastasis [23]. A study by Hong et al., has shown that pancreatic CSCs are consistent with gemcitabine-resistant cells as they are more tumorigenic than gemcitabine sensitive cells [35]. These studies suggest that pancreatic CSCs contribute to chemoresistance as well as metastasis. Further understanding of this process is warranted, as complications associated with tumor metastasis are the major cause of death in pancreatic cancer patients. Expression of adhesion molecules, CRs, and their respective ligands provide opportunities to regulate the metastatic process by selective inhibition. Whether the metastatic process is preventable/blockable in pancreatic cancer is yet to be determined, but interactions of CSCs with surface molecules/receptors may provide clinically relevant targets when considering therapies to inhibit invasion and metastasis of CSCs.

8.3 Pancreatic CSCs and EMT

EMT, epithelial to mesenchymal transition, is a process by which cells lose epithelial characteristics and gain mesenchymal properties (Fig. 8.2). These cells are characterized by an increased potential to invade surrounding tissues as well as enter the circulatory system to seed metastases in distant organs [36–38]. E-cadherin loss is a

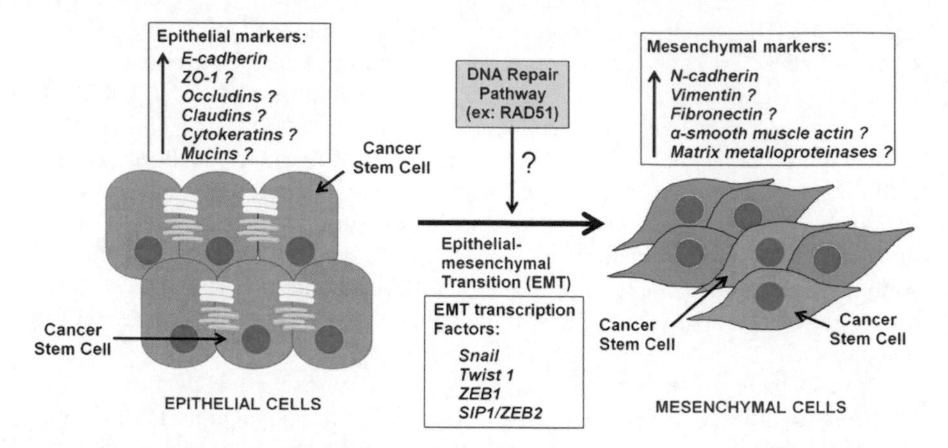

Fig. 8.2 DNA repair pathway and CSCs in EMT. Epithelial cells (*left*) are tightly interconnected by numerous cell-cell interactions. During EMT, these cells lose cell-cell junctions and the actin cytoskeleton is reorganized. E-cadherin repressors induce EMT by regulating expression of genes that transform the cancer cells from epithelial state (*left*) to mesenchymal state (*right*) by suppressing epithelial markers and expressing mesenchymal markers. EMT cancer cells and CSCs have unique roles in the development of tumor metastasis. Many epithelial markers (Ex. ZO-1, occludins, claudins, cytokeratins and mucins) or mesenchymal markers (Ex. Vimentin, fibronectin, α-smooth muscle actin and matrix metalloproteinases) have not yet been characterized in pancreatic CSC biology. The transcription factors which may be involved in the EMT process of pancreatic CSCs are Snail, Twist1, ZEB1 and SIP1/ZEB2. DNA repair pathway genes involved in the acquisition of EMT and CSCs in tumorigenicity are unknown. A potential DNA repair gene involved in pancreatic EMT may be through over expression of RAD51

property associated with these phenotypic changes and is a hallmark of EMT [39]. EMT is considered a prerequisite to metastasis for most cancers, allowing these cells to dissociate from the primary tumor and increase cell motility [40]. Mouse models of pancreatic ductal adenocarcinoma (PDAC) recapitulate this relationship [41]. The relationship between EMT, a proposed mediator of metastatic disease and CSCs has been recently identified in a number of studies [38]. The development of EMT in CSCs is triggered by the interplay of multiple cellular signaling pathways such as Hedgehog, Notch, PDGF, Wnt, TGF-β, Akt, and NF-κB [42–49] demonstrating dysregulation of molecular mechanisms involved in the mesenchymal transition of CSCs. Moreover, Nodal and Activin, members of the transforming growth factor beta (TGF-β) super family, joins a growing list of developmental pathways that are deregulated in pancreatic CSCs [50, 51]. Nodal and its co-receptor Cripto-1 are inappropriately expressed in various human tumors including pancreatic CSCs and may promote tumorigenesis by facilitating EMT or the CSC phenotype.

Studies demonstrating that the ectopic expression of the Snail or Twist transcription factors in immortalized mammary epithelial cells induced EMT [52]. This study also proved the capacity to form mammospheres *in vitro* or in tumors *in vivo* suggesting a link between EMT and CSCs. An association between chemoresistance, the mesenchymal phenotype, and CSCs has been studied and found in pancreatic cancer

[53, 54]. Wang et al., reported that gemcitabine-resistant cells show the acquisition of EMT phenotype by decreasing E-cadherin expression and increasing ZEB and vimentin expression [43]. Recently, it was confirmed that pancreatic cancer cells that are resistant to gemcitabine have CSC features [35, 55]. Pancreatic CSCs showed an up-regulation of sonic hedgehog signaling [56], and targeting hedgehog resulted in the inhibition of CSCs and EMT with decreased expression of Snail and increased expression of E-cadherin leading to the inhibition of invasion and metastasis in pancreatic cancer [57, 58]. Multiple findings show that a number of factors including Twist1, ZEB1, ZEB2, the TGF-β pathway, and microRNAs are able to regulate both EMT and CSC function, which have further strengthened the association between CSCs and EMT [49, 59–61] (Fig. 8.2). Therefore, the biology of CSCs must be determined through precisely studied clinical and laboratory situations, since the molecular mechanisms involved in the mediation of tumor formation and growth are very complex and interrelated. Whether these CSC markers are of functional significance or are markers of convenience remains unknown since we have very little understanding of its significance.

8.4 Pancreatic CSCs and DNA Repair Capacity

Recent evidence indicate the importance of DNA repair in SC maintenance and show that DNA repair genes are highly expressed in SCs. It is now clear that the absence of DNA repair can lead to loss of regulatory pathways with subsequent alterations in gene expression, replication and genomic instability. These occurrences uncover the possibility of loss of SCs in all organ systems. In humans, loss of the SC phenotype is associated with an increase in number of defects in DNA. The DNA repair pathways in which these defects have been identified include MMR, NHEJ, V(D)J end-joining, DSB repair and NER [62]. During SC proliferation, these important defects in the DNA repair pathway imply a necessity for maintenance of DNA repair.

Defects of DNA repair are responsible for several human diseases including SC failure syndromes, aging and cancer [62, 63]. SCs undergo asynchronous DNA synthesis, asymmetric self-renewal and immortal DNA strand co-segregation which prevents accumulation of mutations associated with replication errors or DNA lesions arising from DNA damaging agents [4, 64–66]. The protection of SCs is ensured by various mechanisms, for example, there is an enhancement of DNA repair in normal SCs. Whether this holds the same for CSCs is still being studied. Normal SCs usually protect their genome through enhanced DNA repair unlike in CSCs. When a normal SC transforms and becomes a cancerous SC, it poses a problem for DNA repair as the DNA repair rates are normal, but low proliferation and constitutive activation of the DNA damage checkpoint response confers increased time for lesion removal or bypass before arrival of the replication fork. These types of characteristics may be general to SCs from many cancer types [67].

Knowledge regarding the roles of DNA repair pathways and deficiencies or abnormalities in CSCs affecting the development of various disease processes has increased exponentially over the past few years. The development of cancer risk [68] and therapy outcome [69] have been associated with mutations in DNA repair genes implying that the DNA repair is a double-edged sword in SCs. DNA repair protects normal SCs in both embryonic and adult tissues from genetic damage, thus allowing maintenance of intact genomes into new tissue. Enthusiasm in the analysis or determination of an individual's DNA repair capacity is high because of the importance of DNA repair mechanism in human cancer. A significant increase in gene expression especially genes involved in BRCA1-mediated DNA repair has been observed in a recent study by using a model of invasion to isolate pancreatic tumor initiating CSCs [17]. With respect to BRCA1 in SC fate, studies have shown that loss of BRCA1 results in an accumulation of genetically unstable breast SCs potentially leading to future transforming events [70]. Furthermore, BRCA1 appears to be crucial in mammary gland development and differentiation. Knock-down of its expression has been shown to inhibit mammary epithelial cells from forming acini in 3D cultures [70, 71]. Therefore, BRCA1 appears to be a major stem-cell regulator controlling aspects of development and differentiation, as well as influencing transformation and carcinogenic potential [72]. The significance of functional BRCA1 in SC fate has been established; however, the effects of dysfunctional BRCA1 expression also have implications for anti-cancer therapeutic sensitivity. There is evidence supporting increased sensitization to chemotherapeutic agents when BRCA1 is deficient, as well as when it is exogenously over expressed. Several studies support increased sensitivity to the DNA damaging agents cisplatin, etoposide and bleomycin, in BRCA1-deficient cells, with special emphasis on greater susceptibility to mitomycin C in mouse embryonic SCs [73]. The recent interest in poly ADP ribose polymerase (PARP) inhibitors also applies to BRCA1-deficient cells. Conversely, there is support that over expression of BRCA1 in cancer cells and xenograft models also confers greater sensitivity to cisplatin, as well as lovastatin, an inhibitor of 3-hydroxy-3-methylglutaryl coenzyme A reductase (HMG CoA) [73, 74]. The role of BRCA1 sensitization in transformed SCs could be of great impact when it is taken into consideration that transformed SCs are particularly resistant to chemo- and radio-therapeutic agents leading to relapses, also increasing the growth of tumors. Therefore, novel agents that may sensitize chemo- and radio-therapy are necessary prequels to a strong or in depth research identifying the contribution of DNA repair proteins, with BRCA1 being an example, in resistance of tumors.

8.5 Pancreatic CSCs and RAD51

Even though both NHEJ and HR pathways play an important role in the repair of DSBs, HR is considered an error free pathway for the repair of DSBs and for the maintenance of genome integrity [75–77]. RAD51 is a key protein of *E. coli* RecA ortholog, which plays a major role in the HR repair pathway and is essential for the stability of the genome and a normal cell cycle. The important role of RAD51 in HR is supported by the fact that a knockout mouse of RAD51 shows an early embryonic lethality [78–81]. Principally, higher expression of RAD51 has been observed in various types of human cancers including pancreatic [82–85], implying that increased levels of RAD51 may contribute to increase in tumorigenicity. Over expressed RAD51 has implications for regulation of HR activity and modulation of DNA binding. Despite the crucial role that RAD51 has in repairing damaged DNA, its expression must be tightly regulated. RAD51 must be expressed at required levels needed for proficient execution of DSB repair. Understandably, decreased RAD51 expression results in compromised and inefficient repair of DSBs, which leads to genomic instability. However, over expressed RAD51 leads to heighten HR which causes aberrant recombination [86–88]. This disruption leads to recombination between non-homologous as well as homologous sequences and unequal exchanges of genetic information, often seen between short repetitive DNA sequences (such as *Alu* sites). Furthermore, heightened HR disrupts recombination by causing intra chromosomal crossing-over, and has been correlated with chromosomal anomalies such as translocations and rearrangements [86–88]. The resulting genomic instability from over expressed RAD51 serves as a major precursor in the malignant transformation of normal cells by allowing a mutagenic-permissive cellular environment [86–88]. Furthermore, over expression of RAD51 is correlated with heightened cellular proliferation and chemotherapeutic/radiation insensitivity [84, 85, 89, 90]. And, most important, multiple tumor types and leukemia's express elevated levels of RAD51. Therefore, the over expression of RAD51, and subsequent hyper-HR, could greatly contribute to the development as well as progression of tumors [84, 85, 89, 90]. It is speculated that the hyper-HR of over expressed RAD51 may deregulate cell cycle checkpoints, which may contribute to chemotherapeutic and radiation resistance. It is not presently understood how deregulation of RAD51 occurs after over expression and how the modulators of RAD51 activity are not as efficient as they normally are when RAD51 is present at baseline or normal concentrations. These observations of RAD51 activity that have been noted in a variety of carcinoma cells may influence the progression of CSCs that over express RAD51.

It is appealing to reason that RAD51, a key player in the recombinational repair of damaged DNA might have a key role in the competent DNA repair mechanisms of CSCs. High expression of RAD51 in ALDH$^+$ CSCs of breast cancer cell lines in a recent study suggested the importance of Chk1 dependent HR in DNA repair of CSCs [91]. RAD51 is required for the repair of stalled replication forks during S phase. RAD51 is activated when phosphorylated by the kinase Chk1 [92], which is a serine/threonine kinase that is a member of the ATM-ATR-Chk1 signaling pathway that is crucial in the DNA damage response. The S phase checkpoint, also known as the point in the cell cycle when the integrity of the DNA is confirmed before proceeding with cell division, is controlled by the ATM-ATR-Chk1 signaling pathway [93–95]. This pathway mediates cell cycle arrest to facilitate concomitant resolution of stalled replication forks resulting from single-strand DNA breaks (SSBs) which, at the replication fork resemble DSBs [93–95]. In this instance, when the replication fork incurs DNA damage, this signaling pathway invokes a checkpoint that induces cell cycle arrest and allows for DNA repair [93–95]. Chk1 phosphorylates RAD51 on threonine-309, to activate it to execute repair of the DNA damage that has caused stalling of the replication fork [96]. Therefore, regulation of RAD51 activity and expression appear to be vital in CSCs of varying cell origins. Interestingly, some of the DNA repair genes like *Xpg*, *Ku80*, *Msh2*, and *Rad23b* or *Xrcc1/lig3*, *Ercc2/Xpd* and *Msh2* that are involved in MMR and NHEJ are highly expressed in SCs [97, 98]. It is unknown whether RAD51 is highly expressed or modulated differentially in order to increase the DNA repair efficiency in CSCs by recombination mediators. The role of RAD51 in the acquisition of EMT and CSCs in tumorigenicity is unknown and our recent study showed that over expression of RAD51 is mechanistically linked to chemoresistance and is consistent with the acquisition of an EMT phenotype in pancreatic cancer [66] (Fig. 8.2). This hypothesis must be tested to understand the role of RAD51 in CSCs and to provide insight into considering RAD51 as a potential target for pancreatic cancer therapy.

8.6 Conclusions and Perspectives

Our current understanding of pancreatic cancer progression and metastasis has been revolutionized by studying CSCs, yet many things remain to be elucidated regarding their role. It should be noted that the CSC population itself is a heterogeneous population. Hence, it is important to understand and study the genuine nature of the CSC population during cancer progression. It is important to purify and evaluate CSC populations. We are now realizing that the cellular hierarchy which is defined by the CSC theory is more complex than originally thought. It is not enough to identify a pure CSC population through a single marker as CSCs can express or lack the conventional CSC marker CD133 and still retain the functional characteristics that define a CSC as demonstrated in glioblastoma [99]. Moreover, it is not clear whether known CSC populations have distinct progressive, invasive or metastatic potential. Therefore, in addition to CSCs defining characteristic of tumorigenicity, a number of properties such as migratory and invasive potential are now attributed to CSCs that suggest a primary role in disease relapse and progression.

Studying the effect of CSCs in relation to tumorigenicity, metastasis and therapeutic resistance properties of tumors is a reasonable step on our path in investigating the mechanisms by which CSCs are distinct from the remaining cancer cell population. A fascinating insight into failure of available therapies in pancreatic cancer is offered by studying CSCs. Many lines of evidences suggest that CSCs show increased resistance to chemo- and radiation therapies in pancreatic cancer, but the mechanism is not completely known. Identification of CSCs that exhibit these properties requires focused research to overcome these resistance mechanism(s). Increasing evidence also suggests that a tumors metastatic potential in CSCs may be evident from the tumor microenvironment in which they exist. Additional research in the pancreatic CSCs is needed to understand the molecular mechanisms regulating self-renewal and tumor metastasis. Recent work by Hermann et al., found that a small subpopulation of pancreatic CSCs (CD133^{+}CXCR4^{+}) were able to strongly metastasize, and that if CXCR4 was inhibited, the cells were unable to metastasize [23]. These types of studies may yield important insights into molecular therapeutic approaches and outcome results in pancreatic cancer patient care.

Whether, the existence of cancer prone SCs with defects in DNA repair that can be identified before they become malignant is not clear. Understanding the transition towards a malignant transformation is one of the most important research areas of SCs responsible for cancer progression and metastasis. The DNA repair pathway appears to be tightly connected between normal and CSCs and similarly the SC phenotypes are also connected between normal and cancer cells. Therefore, identification and development of biomarkers and suitable assays are important in identifying DNA repair anomalies in clinical samples. The DNA repair protein RAD51 may be a key component in the etiology of CSCs. And, as more investigations of overexpressed RAD51 reveal its mechanistic role in influencing malignant transformation, increased cell proliferation, and resistance to both chemotherapeutics and radiation, RAD51 could potentially serve as a vital future biomarker, as well as chemotherapeutic target. Comprehending this progression remains an exciting field of investigation in pancreatic CSC biology.

Acknowledgments We thank Drs. Nipun Merchant, Elaine Hurt, Stephanie Cabarcas, Lesley Mathews and Michael Van Saun for their useful comments and stimulating discussions during the preparation of this manuscript.

References

1. Parkin DM, Bray F, Ferlay J et al (2005) Global cancer statistics, 2002. CA Cancer J Clin 55:74–108
2. Clevers H (2011) The cancer stem cell: premises, promises and challenges. Nat Med 17:313–319
3. Clarke MF, Dick JE, Dirks PB et al (2006) Cancer stem cells—perspectives on current status and future directions: AACR Workshop on cancer stem cells. Cancer Res 66:9339–9344
4. Tang C, Ang BT, Pervaiz S (2007) Cancer stem cell: target for anti-cancer therapy. FASEB J 21:3777–3785
5. Ma S, Chan KW, Hu L et al (2007) Identification and characterization of tumorigenic liver cancer stem/progenitor cells. Gastroenterology 132:2542–2556
6. O'Brien CA, Pollett A, Gallinger S et al (2007) A human colon cancer cell capable of initiating tumour growth in immunodeficient mice. Nature 445:106–110
7. Kim CF, Jackson EL, Woolfenden AE et al (2005) Identification of bronchioalveolar stem cells in normal lung and lung cancer. Cell 121:823–835
8. Lee CJ, Dosch J, Simeone DM (2008) Pancreatic cancer stem cells. J Clin Oncol 26:2806–2812
9. Hanahan D, Weinberg RA (2000) The hallmarks of cancer. Cell 100:57–70
10. Feinberg AP, Ohlsson R, Henikoff S (2006) The epigenetic progenitor origin of human cancer. Nat Rev Genet 7:21–33
11. Hastings PJ, Lupski JR, Rosenberg SM et al (2009) Mechanisms of change in gene copy number. Nat Rev Genet 10:551–564
12. Stratton MR, Campbell PJ, Futreal PA (2009) The cancer genome. Nature 458:719–724
13. Hoeijmakers JH (2001) Genome maintenance mechanisms for preventing cancer. Nature 411:366–374
14. O'Driscoll M, Jeggo PA (2006) The role of double-strand break repair—insights from human genetics. Nat Rev Genet 7:45–54

15. Wyman C, Kanaar R (2006) DNA double-strand break repair: all's well that ends well. Annu Rev Genet 40:363–383
16. Heyer WD, Ehmsen KT, Liu J (2010) Regulation of homologous recombination in eukaryotes. Annu Rev Genet 44:113–139
17. Mathews LA, Cabarcas SM, Farrar WL (2011) DNA repair: the culprit for tumor-initiating cell survival? Cancer Metastasis Rev 30:185–197
18. Ralhan R, Kaur J, Kreienberg R et al (2007) Links between DNA double strand break repair and breast cancer: accumulating evidence from both familial and nonfamilial cases. Cancer Lett 248:1–17
19. Sarasin A, Kauffmann A (2008) Overexpression of DNA repair genes is associated with metastasis: a new hypothesis. Mutat Res 659:49–55
20. Adhikari AS, Agarwal N, Iwakuma T (2011) Metastatic potential of tumor-initiating cells in solid tumors. Front Biosci 16:1927–1938
21. Zischek C, Niess H, Ischenko I et al (2009) Targeting tumor stroma using engineered mesenchymal stem cells reduces the growth of pancreatic carcinoma. Ann Surg 250:747–753
22. Li C, Heidt DG, Dalerba P et al (2007) Identification of pancreatic cancer stem cells. Cancer Res 67:1030–1037
23. Hermann PC, Huber SL, Herrler T et al (2007) Distinct populations of cancer stem cells determine tumor growth and metastatic activity in human pancreatic cancer. Cell Stem Cell 1:313–323
24. Rasheed ZA, Yang J, Wang Q et al (2010) Prognostic significance of tumorigenic cells with mesenchymal features in pancreatic adenocarcinoma. J Natl Cancer Inst 102:340–351
25. Visvader JE, Lindeman GJ (2008) Cancer stem cells in solid tumours: accumulating evidence and unresolved questions. Nat Rev Cancer 8:755–768
26. Munz M, Kieu C, Mack B et al (2004) The carcinoma-associated antigen EpCAM upregulates c-myc and induces cell proliferation. Oncogene 23:5748–5758
27. Thayer SP, di Magliano MP, Heiser PW et al (2003) Hedgehog is an early and late mediator of pancreatic cancer tumorigenesis. Nature 425:851–856
28. Narducci MG, Scala E, Bresin A et al (2006) Skin homing of Sezary cells involves SDF-1-CXCR4 signaling and down-regulation of CD26/dipeptidylpeptidase IV. Blood 107:1108–1115
29. Klein RS, Rubin JB, Gibson HD et al (2001) SDF-1 alpha induces chemotaxis and enhances Sonic hedgehog-induced proliferation of cerebellar granule cells. Development 128:1971–1981
30. Doitsidou M, Reichman-Fried M, Stebler J et al (2002) Guidance of primordial germ cell migration by the chemokine SDF-1. Cell 111:647–659
31. Aiuti A, Webb IJ, Bleul C et al (1997) The chemokine SDF-1 is a chemoattractant for human $CD34^{+}$ hematopoietic progenitor cells and provides a new mechanism to explain the mobilization of $CD34^{+}$ progenitors to peripheral blood. J Exp Med 185:111–120
32. Talmadge JE, Fidler IJ (2010) AACR centennial series: the biology of cancer metastasis: historical perspective. Cancer Res 70:5649–5669
33. Mimeault M, Johansson SL, Senapati S et al (2010) MUC4 down-regulation reverses chemoresistance of pancreatic cancer stem/progenitor cells and their progenies. Cancer Lett 295:69–84
34. Yao J, Cai HH, Wei JS et al (2010) Side population in the pancreatic cancer cell lines SW1990 and CFPAC-1 is enriched with cancer stem-like cells. Oncol Rep 23:1375–1382
35. Hong SP, Wen J, Bang S et al (2009) CD44-positive cells are responsible for gemcitabine resistance in pancreatic cancer cells. Int J Cancer 125:2323–2331
36. Yang J, Weinberg RA (2008) Epithelial-mesenchymal transition: at the crossroads of development and tumor metastasis. Dev Cell 14:818–829
37. Thiery JP, Sleeman JP (2006) Complex networks orchestrate epithelial-mesenchymal transitions. Nat Rev Mol Cell Biol 7:131–142
38. Thiery JP (2002) Epithelial-mesenchymal transitions in tumour progression. Nat Rev Cancer 2:442–454
39. Kang Y, Massague J (2004) Epithelial-mesenchymal transitions: twist in development and metastasis. Cell 118:277–279

40. Wels J, Kaplan RN, Rafii S et al (2008) Migratory neighbors and distant invaders: tumor-associated niche cells. Genes Dev 22:559–574
41. von Burstin J, Eser S, Paul MC et al (2009) E-cadherin regulates metastasis of pancreatic cancer in vivo and is suppressed by a SNAIL/HDAC1/HDAC2 repressor complex. Gastroenterology 137;361–371, 371 e361–365
42. Singh A, Greninger P, Rhodes D et al (2009) A gene expression signature associated with "K-Ras addiction" reveals regulators of EMT and tumor cell survival. Cancer Cell 15:489–500
43. Wang Z, Li Y, Kong D et al (2009) Acquisition of epithelial-mesenchymal transition phenotype of gemcitabine-resistant pancreatic cancer cells is linked with activation of the notch signaling pathway. Cancer Res 69:2400–2407
44. Huber MA, Azoitei N, Baumann B et al (2004) NF-kappaB is essential for epithelial-mesenchymal transition and metastasis in a model of breast cancer progression. J Clin Invest 114:569–581
45. Moustakas A, Heldin CH (2007) Signaling networks guiding epithelial-mesenchymal transitions during embryogenesis and cancer progression. Cancer Sci 98:1512–1520
46. Zavadil J, Bottinger EP (2005) TGF-beta and epithelial-to-mesenchymal transitions. Oncogene 24:5764–5774
47. Timmerman LA, Grego-Bessa J, Raya A et al (2004) Notch promotes epithelial-mesenchymal transition during cardiac development and oncogenic transformation. Genes Dev 18:99–115
48. Zavadil J, Cermak L, Soto-Nieves N et al (2004) Integration of TGF-beta/Smad and Jagged1/Notch signalling in epithelial-to-mesenchymal transition. EMBO J 23:1155–1165
49. Sarkar FH, Li Y, Wang Z et al (2009) Pancreatic cancer stem cells and EMT in drug resistance and metastasis. Minerva Chir 64:489–500
50. Bianco C, Rangel MC, Castro NP et al (2010) Role of Cripto-1 in stem cell maintenance and malignant progression. Am J Pathol 177:532–540
51. Lonardo E, Hermann PC, Mueller MT et al (2011) Nodal/Activin signaling drives self-renewal and tumorigenicity of pancreatic cancer stem cells and provides a target for combined drug therapy. Cell Stem Cell 9:433–446
52. Mani SA, Guo W, Liao MJ et al (2008) The epithelial-mesenchymal transition generates cells with properties of stem cells. Cell 133:704–715
53. Wang Z, Li Y, Ahmad A et al (2011) Pancreatic cancer: understanding and overcoming chemoresistance. Nat Rev Gastroenterol Hepatol 8:27–33
54. Shah AN, Summy JM, Zhang J et al (2007) Development and characterization of gemcitabine-resistant pancreatic tumor cells. Ann Surg Oncol 14:3629–3637
55. Wang Z, Ahmad A, Li Y et al (2011) Targeting notch to eradicate pancreatic cancer stem cells for cancer therapy. Anticancer Res 31:1105–1113
56. Li C, Lee CJ, Simeone DM (2009) Identification of human pancreatic cancer stem cells. Methods Mol Biol 568:161–173
57. Feldmann G, Fendrich V, McGovern K et al (2008) An orally bioavailable small-molecule inhibitor of Hedgehog signaling inhibits tumor initiation and metastasis in pancreatic cancer. Mol Cancer Ther 7:2725–2735
58. Feldmann G, Dhara S, Fendrich V et al (2007) Blockade of hedgehog signaling inhibits pancreatic cancer invasion and metastases: a new paradigm for combination therapy in solid cancers. Cancer Res 67:2187–2196
59. Wellner U, Schubert J, Burk UC et al (2009) The EMT-activator ZEB1 promotes tumorigenicity by repressing stemness-inhibiting microRNAs. Nat Cell Biol 11:1487–1495
60. Shimono Y, Zabala M, Cho RW et al (2009) Downregulation of miRNA-200c links breast cancer stem cells with normal stem cells. Cell 138:592–603
61. Brabletz S, Brabletz T (2010) The ZEB/miR-200 feedback loop—a motor of cellular plasticity in development and cancer? EMBO Rep 11:670–677
62. Gerson SL, Keynon J, Qing YL (2008) DNA repair: an essential role in stem cell maintenance. Blood Cell Mol Dis 40:267–268
63. Tutt A, Ashworth A (2002) The relationship between the roles of BRCA genes in DNA repair and cancer predisposition. Trends Mol Med 8:571–576

64. Croker AK, Allan AL (2008) Cancer stem cells: implications for the progression and treatment of metastatic disease. J Cell Mol Med 12:374–390
65. Wicha MS, Liu S, Dontu G (2006) Cancer stem cells: an old idea—a paradigm shift. Cancer Res 66:1883–1890; discussion 1895–1886
66. Nagathihalli NS, Nagaraju G (2011) RAD51 as a potential biomarker and therapeutic target for pancreatic cancer. Biochim Biophys Acta 1816:209–218
67. Viale A, De Franco F, Orleth A et al (2009) Cell-cycle restriction limits DNA damage and maintains self-renewal of leukaemia stem cells. Nature 457:51–56
68. Negrini S, Gorgoulis VG, Halazonetis TD (2010) Genomic instability—an evolving hallmark of cancer. Nat Rev Mol Cell Biol 11:220–228
69. Olaussen KA, Dunant A, Fouret P et al (2006) DNA repair by ERCC1 in non-small-cell lung cancer and cisplatin-based adjuvant chemotherapy. N Engl J Med 355:983–991
70. Liu S, Ginestier C, Charafe-Jauffret E et al (2008) BRCA1 regulates human mammary stem/progenitor cell fate. Proc Natl Acad Sci USA 105:1680–1685
71. Molyneux G, Geyer FC, Magnay FA et al (2010) BRCA1 basal-like breast cancers originate from luminal epithelial progenitors and not from basal stem cells. Cell Stem Cell 7:403–417
72. Smalley MJ, Reis-Filho JS, Ashworth A (2008) BRCA1 and stem cells: tumour typecasting. Nat Cell Biol 10:377–379
73. James CR, Quinn JE, Mullan PB et al (2007) BRCA1, a potential predictive biomarker in the treatment of breast cancer. Oncologist 12:142–150
74. Yu X, Luo Y, Zhou Y et al (2008) BRCA1 overexpression sensitizes cancer cells to lovastatin via regulation of cyclin D1-CDK4-p21WAF1/CIP1 pathway: analyses using a breast cancer cell line and tumoral xenograft model. Int J Oncol 33:555–563
75. Moynahan ME, Jasin M (2010) Mitotic homologous recombination maintains genomic stability and suppresses tumorigenesis. Nat Rev Mol Cell Biol 11:196–207
76. Nagaraju G, Scully R (2007) Minding the gap: the underground functions of BRCA1 and BRCA2 at stalled replication forks. DNA Repair (Amst) 6:1018–1031
77. Sung P, Klein H (2006) Mechanism of homologous recombination: mediators and helicases take on regulatory functions. Nat Rev Mol Cell Biol 7:739–750
78. San Filippo J, Sung P, Klein H (2008) Mechanism of eukaryotic homologous recombination. Annu Rev Biochem 77:229–257
79. Paques F, Haber JE (1999) Multiple pathways of recombination induced by double-strand breaks in Saccharomyces cerevisiae. Microbiol Mol Biol Rev 63:349–404
80. Symington LS (2002) Role of RAD52 epistasis group genes in homologous recombination and double-strand break repair. Microbiol Mol Biol Rev 66:630–670 (table of contents)
81. Tsuzuki T, Fujii Y, Sakumi K et al (1996) Targeted disruption of the Rad51 gene leads to lethality in embryonic mice. Proc Natl Acad Sci USA 93:6236–6240
82. Maacke H, Jost K, Opitz S et al (2000) DNA repair and recombination factor Rad51 is over-expressed in human pancreatic adenocarcinoma. Oncogene 19:2791–2795
83. Richardson C (2005) RAD51, genomic stability, and tumorigenesis. Cancer Lett 218:127–139
84. Klein HL (2008) The consequences of Rad51 overexpression for normal and tumor cells. DNA Repair (Amst) 7:686–693
85. Henning W, Sturzbecher HW (2003) Homologous recombination and cell cycle checkpoints: Rad51 in tumour progression and therapy resistance. Toxicology 193:91–109
86. Vispe S, Cazaux C, Lesca C et al (1998) Overexpression of Rad51 protein stimulates homologous recombination and increases resistance of mammalian cells to ionizing radiation. Nucleic Acids Res 26:2859–2864
87. Richardson C, Jasin M (2000) Frequent chromosomal translocations induced by DNA double-strand breaks. Nature 405:697–700
88. Richardson C, Stark JM, Ommundsen M et al (2004) Rad51 overexpression promotes alternative double-strand break repair pathways and genome instability. Oncogene 23:546–553
89. Connell PP, Jayathilaka K, Haraf DJ et al (2006) Pilot study examining tumor expression of RAD51 and clinical outcomes in human head cancers. Int J Oncol 28:1113–1119

90. Hannay JA, Liu J, Zhu QS et al (2007) Rad51 overexpression contributes to chemoresistance in human soft tissue sarcoma cells: a role for p53/activator protein 2 transcriptional regulation. Mol Cancer Ther 6:1650–1660
91. Charafe-Jauffret E, Ginestier C, Iovino F et al (2009) Breast cancer cell lines contain functional cancer stem cells with metastatic capacity and a distinct molecular signature. Cancer Res 69:1302–1313
92. Sorensen CS, Hansen LT, Dziegielewski J et al (2005) The cell-cycle checkpoint kinase Chk1 is required for mammalian homologous recombination repair. Nat Cell Biol 7:195–201
93. Liu Q, Guntuku S, Cui XS et al (2000) Chk1 is an essential kinase that is regulated by Atr and required for the G(2)/M DNA damage checkpoint. Genes Dev 14:1448–1459
94. Takai H, Tominaga K, Motoyama N et al (2000) Aberrant cell cycle checkpoint function and early embryonic death in Chk1(−/–) mice. Genes Dev 14:1439–1447
95. Bartek J, Lukas J (2003) Chk1 and Chk2 kinases in checkpoint control and cancer. Cancer Cell 3:421–429
96. Hurley PJ, Bunz F (2007) ATM and ATR: components of an integrated circuit. Cell Cycle 6:414–417
97. Ramalho-Santos M, Yoon S, Matsuzaki Y et al (2002) "Stemness": transcriptional profiling of embryonic and adult stem cells. Science 298:597–600
98. Ivanova NB, Dimos JT, Schaniel C et al (2002) A stem cell molecular signature. Science 298:601–604
99. Chen R, Nishimura MC, Bumbaca SM et al (2010) A hierarchy of self-renewing tumor-initiating cell types in glioblastoma. Cancer Cell 17:362–375

Chapter 9
Targeting Cancer Stem Cell Efficient DNA Repair Pathways: Screening for New Therapeutics

Lesley A. Mathews, Francesco Crea and Marc Ferrer

Abstract The existence of 'cancer stem cells (CSCs)' has been a topic of vigorous discussion for the last few years within the field of cancer biology. Continuous characterization of tumor cells has lead to an abundance of data supporting the existence of cell populations with stem cell characteristics, including self-renewal and expression of stem cell markers. There is also evidence suggesting that these cells are responsible for chemo- and radio-resistance and are the initiation point for metastasis, cancer recurrence, and ultimately patient demise. Therefore, finding new drugs that induce cancer stem cell death are of high interest as new therapies for cancer. Gene expression arrays, functional genomics screens with siRNA, as well as screening of small molecule libraries are approaches being used to better understand the cellular pathways that are critical for cancer stem cell survival. Finding drugs that target these pathways in cancer stem cells could represent novel therapies for cancer, in particular for the prevention of metastasis and recurrence. Recent data shows that DNA repair genes are upregulated in pancreatic cancer stem cells, thus providing increased genomic stability and resistance to cell death upon treatment with DNA damaging agents such as gemcitibine. Here we review how a higher efficiency of DNA repair in cancer stem cells can be leveraged therapeutically, and discuss how small molecule screening approaches using stem cells are being used to find new potential therapies that result in terminal differentiation or cell death of cancer stem cells, both as single agents or in combination with other chemotherapeutics.

L. A. Mathews (✉) · M. Ferrer
Division of Preclinical Innovation, National Center for Advancing
Translational Sciences, National Institutes of Health, Rockville, MD 20850, USA
e-mail: mathewsla@mail.nih.gov

F. Crea
Department of Internal Medicine, Division of Pharmacology,
Pisa Medical School, Pisa, Italy

L. A. Mathews et al. (eds.), *DNA Repair of Cancer Stem Cells,*
DOI 10.1007/978-94-007-4590-2_9, © Springer Science+Business Media Dordrecht 2013

9.1 Current Therapeutics Targeting DNA Repair Pathways

DNA-damaging agents are among the most effective classes of drugs for the treatment of cancer. However, cancer stem cell (CSC) populations are very resistant to DNA-damaging agents, in part due to their robust DNA repair mechanisms. Moreover, CSCs lower proliferation rates and constitutive activation of the checkpoint responses allows them to have more time for DNA repair. This reliance of cancer stem cells on DNA repair pathways ensures the genome stability necessary for their self-renewal and provides a possible intervention point for development of treatments that prevent their survival. In this regard, one class of inhibitors, DNA repair dependent-checkpoint inhibitors have been tested and shown to increase the sensitivity of cancer stem cells to radiation therapy (reviewed in [1]).

Checkpoint kinases coordinate activation of the DNA repair machinery and are critical to the DNA damage response by inducing cell cycle arrest, and activating effective DNA repair or induction of cell death if the damage exceeds the capacity of the cell to repair efficiently. AZD7762 is an ATP-competitive checkpoint kinase inhibitor that assists DNA-damaging drugs by blocking the checkpoint response (reviewed in [1]). A recent paper by Gallmeier et al. demonstrated that inhibition of ataxia telangiectasia (ATM)-using CHK1 inhibitor SB218078 abrogates the in vitro and in vivo tumorigenicity of human colon cancer cells through depletion of the CD133(+) tumor-initiating cell fraction [2]. Many other checkpoint kinase inhibitors against CHK1 and CHK2 have been developed and are summarized in a recent review by Garrett and Collins [3]. Although CHK inhibitors have only recently begun to show efficacy at inhibiting CSC growth, future research involving combination therapies hold tremendous promise for eradicating these cells. In this regard, research is now being conducted where the CHK inhibitors are being combined with other chemotherapeutics such as topoisomearse I and II inhibitors, platinum based drugs such as cisplatin, anti-metabolites, mictrotubule-targeting agents and other molecular target agents such as poly (ADP-ribose) polymerase (PARP) inhibitors [3]. Identification of such therapeutic combinations is a slow process, limited in practice to testing a small number of combinations. The use of high throughput screening technologies should help in systematically testing larger number of combinations in cellular assay systems [4, 5], which can then be further validated in in vivo animal models before being tested in the clinic.

Another DNA repair pathway targeted for the killing of CSCs is the repair of the O^6-alkylguanine modified base in DNA, which is regulated by the O^6-alkylguanine DNA methyltransferase (AGT), also called MGMT. MGMT was actually one of the first DNA repair targets identified for cancer therapeutics because its elevated expression levels correlate with drug resistance ([6], reviewed in [7]). It was postulated that MGMT inactivation in tumor cells would have a sensitizing effect to chemotherapeutic agents that generated lesions at the O^6 positions of guanine [7]. In this regard, expression of MGMT in brain cancer cells neutralizes the cytotoxic effect of alkylating agents such as temozolomide (TMZ). A recent study also demonstrated that compared to established glioma cell lines, neurosphere-forming glioma initiating cells (GICs) expressed higher levels of MGMT [8] and when GICs were transduced with an shRNA to MGMT the cells could be sensitized to TMZ treatment by decreasing both their ability to undergo DNA repair and efflux of the drug [8]. Another drug, O^6-benzyguanine (BG) acts as a pseudo-substrate and renders the AGT inactive to repair the DNA damage in tumor cells, and recently it has been shown that in head and neck tumor cells pretreated with BG results in a twofold decrease in the ED_{50} of cisplatin together with an increase in apoptosis and DNA damage ([6], reviewed in [7]).

Additionally, glycosylases are enzymes which remove a single damaged DNA base in need of repair. After base removal, the abasic sites are processed by a protein called APE1/Ref-1. APE1/Ref-1 is essential for base excision repair (BER) and its expression is altered in a number of cancers including prostate, ovarian, cervical and colon ([9–11], reviewed in [7]). In addition, APE1/Ref-1 also stimulates the DNA binding of transcription factors known to regulate cancer such as AP-1 (Fos/Jun), NFκβ, HIF-1α, CREB and p53 (reviewed in [7]). Currently, three compounds have been found to target APE1/Ref-1 activity, including methoxyamine (MX), 7-Nitroindole-2-Carboxylic Acid (NCA) and lucanthone [7]. These compounds synergize with TMZ to kill cancer cells. However, the exact target and mechanism of action of these compounds remains largely unknown.

Finally, non-homologous end-joining (NHEJ) is one of the mechanisms used for double strand break repair (DSB). It takes place during G1/S, requires little or no sequence homology for efficient repair, and can be error-free or error-prone depending on the type of ends that are present at the site of the DSB. During NHEJ, the KU70 and KU80 proteins are recruited to the damaged DNA site, followed by the recruitment and activation of the DNA-protein kinase DNA-PKC, resulting in subsequent activation of XRCC4 and DNA ligase IV (LIG4) [12]. A recent study showed that when DNA-PKC levels were decreased with short hairpin RNA, GICs were radiosensitized and underwent autophagy compared to cells expressing much higher levels of the enzyme [13]. A number of inhibitors are being developed to target DNA-PK including NU7026 [14], IC87361 [15] and IC87102, all of which have been shown to sensitize tumor microvasculature and tumor cells to irradiation (reviewed in [7]). Another mechanism of DSB repair is homologous recombination (HR). HR uses thousands of bases of sequence homology either from a sister chromatid or a homologous chromosome during S/G2 phases and is the most error-free method of repair [16]. A key regulator in mediating which DSB pathway a cell chooses to repair damaged DNA is the multifunctional protein BRCA1 (as reviewed in [17]). BRCA1 preferentially channels DSB repair into HR rather than NHEJ and the process is started by a protein complex containing MRE11, RAD50 and NBS1 termed the MRN complex. RAD51 forms a nucleoprotein filament and catalyzes homologous pairing and strand exchange with the assistance of BRCA2. The MRN complex then recruits ATM, which is a key regulator of HR and the cell-cycle checkpoints. A number of inhibitors have been identified that inhibit ATM such as Wortmannin, LY2940002 and caffeine, however a number of these have been shown to lack selectivity and inhibit other kinases, including PI3K. (reviewed in [7]). Cells which are deficient in BRCA1 and BRCA2, however, are highly sensitive to inhibitors poly (ADP) ribose polymerase-1 (PARP-1) (reviewed in [7]), a protein which interacts with several base excision repair scaffolding proteins. The cells that are deficient in BRCA1 and PARP-1 are not able to initiate homologous recombination and repair the breaks in the DNA. In addition to their enhanced cytotoxic effect in BRAC1 deficient cells, PARP-1 inhibitors also demonstrate sensitization to alkylating agents such as N-methyl-N-nitrosourea (MNU) and chemotherapeutics such at TMZ. In the hepatocellular carcinoma cell (HCC) line Hep-12, which express highlevels of PARP-1, unlike Hep-11 cells, inhibition of PARP-1 potentiated the sensitivity to hydroxycamptothecin (HCPT). According to the authors, this indicates that a large population of the recurrent HCC-derived Hep-12 cells were tumor-initiating cells and that elevated expression of PARP-1 was related to their resistance to HCPT.

Overall, there are a handful of drugs currently available to treat cancer by affecting their DNA repair pathways and some have initially been tested in CSC models, although it is still not clear how these drugs affect the CSC pool. Additional studies will need to test whether these or new drugs which target the DNA repair machinery can as single agents or in combination with other chemotherapeutics and if they can efficiently kill the highly aggressive CSC populations. One approach to search for new compounds that target CSCs utilizes high throughput drug screening.

9.2 High Throughput Screening (HTS) for New CSC Killing Agents

High throughput drug screening (HTS) for lead generation has been used for many years in both pharmaceutical companies, the biotechnology industry and academic institutions, to a lesser extent, to identify new drug candidates. Large collections of compounds (>100,000 compounds) are now routinely screened in 384- or 1536-well microplate format using biochemical or cell-based assays that tend to be addition-only to simplify implementation in automated screening systems. Most cell-based assays used in HTS rely on cell lines engineered with functional reporters, such as GFP or Luciferase, or metabolic readouts of cell viability, such as measurements of ATP levels. Smaller collections (<100,000) of compounds can be practically screened using assays that provide more detailed information on individual cellular events, so called high content screening (HCS) assays. These assays rely on the use of cell lines with fluorescently-tagged proteins or immunohistochemistry-based staining of endogenous proteins. Because of the more complex protocols required to automate these assays, these high content screens are normally implemented in 96- or 384-well format, and therefore have lower throughput. Furthermore, high content screens require significant investments in cell imaging readers and informatics support for image analysis, processing and storage. Therefore HCS, although very powerful, it is not widely used as a screen of a large number of samples. Additional technologies such as flow cytometry and gene expression that provide information-rich assays and are of interest for screening with stem cells. However, technical limitations in

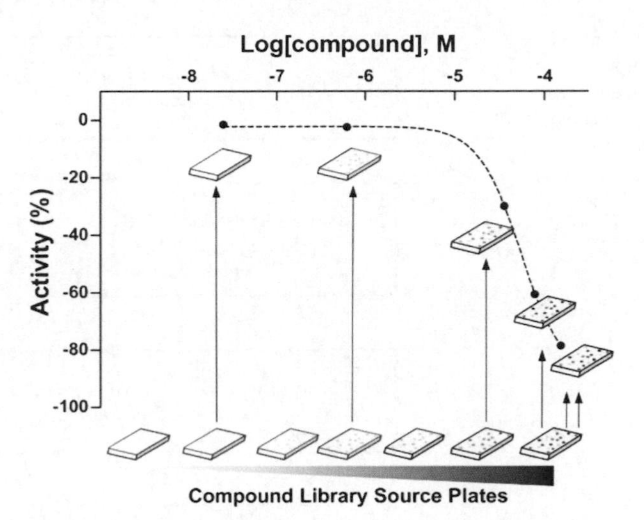

Fig. 9.1 Use of an asymmetric titration series. *Arrows* represent the pin-transfer of compound from the source plates available for testing (*below*) to the assay plates used in qHTS (*above*). In this example, the highest concentration point is customized to contain a twofold higher compound concentration achieved by a double pin-transfer (*dual arrows*) of solution out of the library plate containing the highest available concentration

sample preparation, slow readouts and data analysis, make their use challenging for screening even a small number of compounds (<1000 compounds).

In general, compounds to be screened are prepared as solutions in dimethylsulfoxide (DMSO) and added directly as a DMSO solution or by intermediate dilution in assay buffer to the assay solution. In order to contain cost and resources used, compounds are normally screened at one concentration (~10 uM) and limited to one replicate, although smaller collections (<10,000 compounds) might be tested in replicates. When assays can be miniaturized to 1536-well microplate format, it is possible to implement dose response-based HTS (so called quantitative HTS or qHTS) ([18]. Such an approach has been systematically used at the former NIH Chemical Genomics Center (now under the new National Center for Advancing Translational Sciences or NCATS), and has proven to be advantageous in finding weaker, but biologically relevant active compounds from assays that measured activity of targets traditionally not considered druggable [19] (Figs. 9.1, 9.2 and 9.3).

Currently, the majority of HTS is implemented by testing one unique compound per assay well. It would be desirable to be able to screen a large number of compounds in combination to determine potential additive or synergistic effects of compounds in an assay [5]. The practical implementation of such combination screens is not trivial because the number of pair-wise combinations increase exponentially (e.g., for a 100 compounds generate ~5000 pair wise combinations; 1000 compounds would generate ~500,000) so it would require significant automation capabilities. Current compound plating technologies also limit the number of compound combinations that can be practically prepared in a timely manner. However, by combining

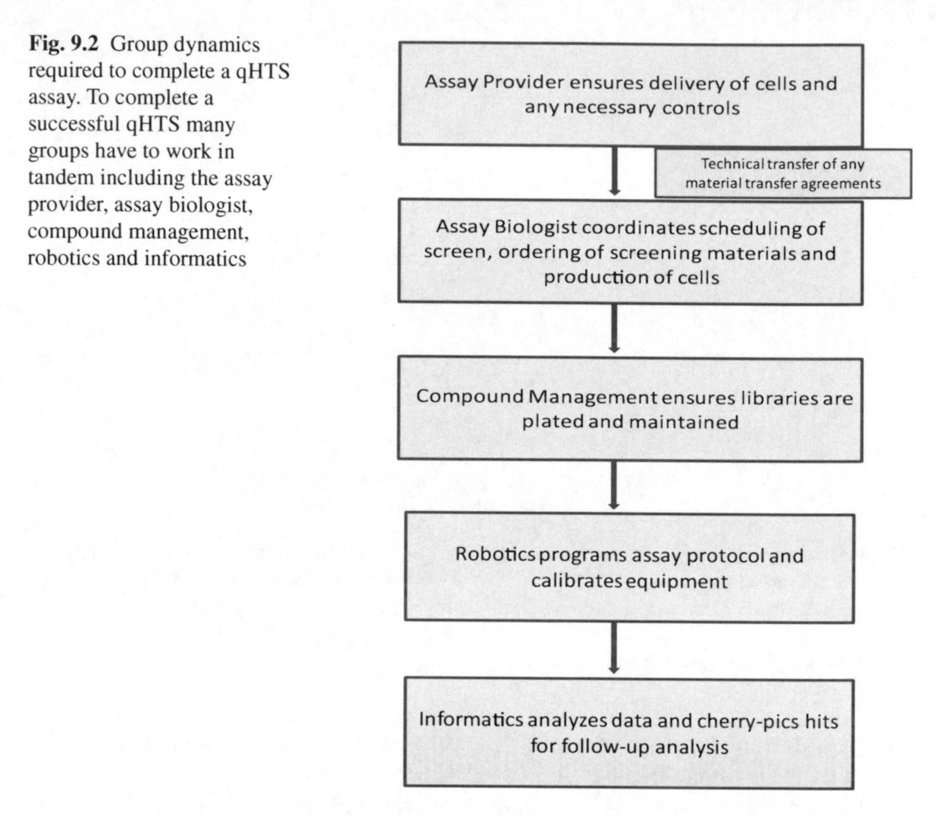

Fig. 9.2 Group dynamics required to complete a qHTS assay. To complete a successful qHTS many groups have to work in tandem including the assay provider, assay biologist, compound management, robotics and informatics

1536-well microplate assay formats with new acoustic-based liquid dispensers, clever compound pooling strategies, and compound plating software, it is now possible to screen pair-wise combinations of up to a thousand of compounds ([20] DPI, NCATS, unpublished data). We expect that this new screening paradigm will enable the systematic testing of a large number of combinations of clinically approved compounds for quick validation in animal models and clinic.

9.3 High Throughput Screening Using Stem Cells

The use of stem cells in HTS has been limited due to the technical challenge of culturing these types of cells, and maintaining homogeneity during scaling up of cell production necessary for the screening process. A list of HTS assays published using stem cells is shown in Tables 9.1 and 9.2. One of the first HTS assays published with stem cells utilized mouse embryonic stem cells (ESCs) containing a stable GFP reporter for the stem cell gene Oct4 [21]. This study highlighted the technical challenges of screening with these cells because when the cells were removed from a feeder cell layer and the stem cell agent LIF was depleted from the media, the cells significantly lose expression of GFP and their compact-colony morphology.

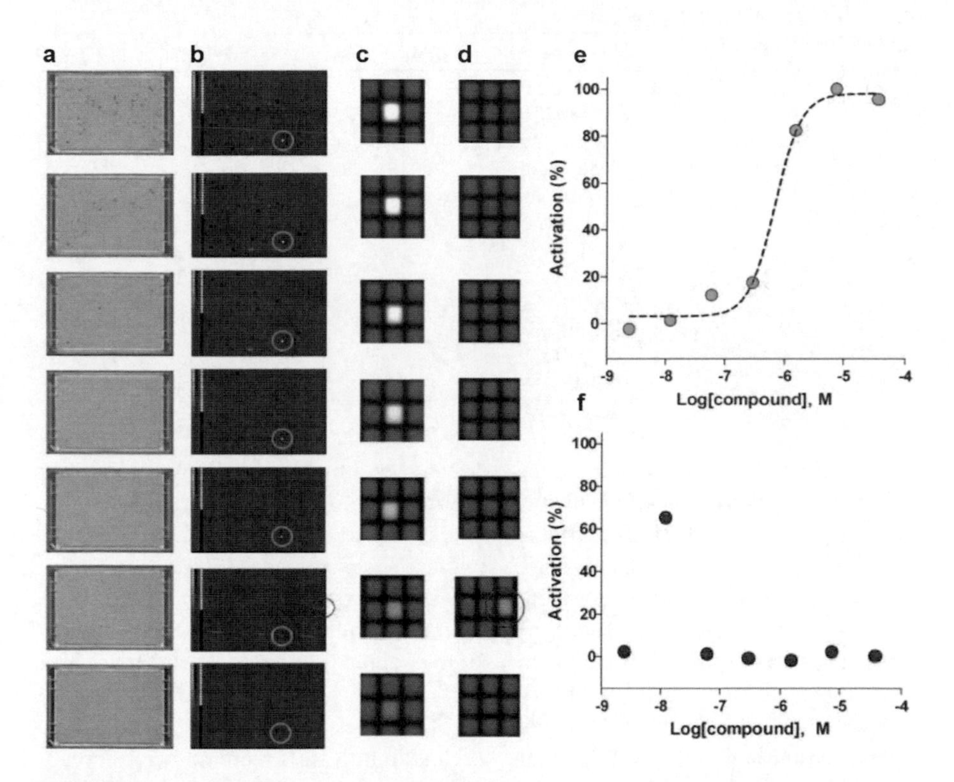

Fig. 9.3 Representative ViewLux data from an inter-plate titration series in qHTS. **a** A library set of seven vertically titrated 1,536-well compound plates arranged from high to low concentration, top to bottom. **b** CCD-based plate-reader images showing an assay response (wells circled in *green*) to increasing compound concentration in a vertically-titrated concentration plate series. A single high-response well (circled in *red*) is observed that does not show a concentration response curve (e.g., a false positive). **c** Magnified images of wells exhibiting concentration response curve. **d** Magnified images of wells in the same region as the false positive well. **e** and **f** Concentration response curves derived from the samples indicated in (**c**) and (**d**), respectively. Curve-fitting software recognizes data from (**d**) as a flat response with a single-point outlier and classifies this as inactive, thus eliminating the sample from unnecessary follow-up analysis

A FACS based screen in 384-well plate format using the same mouse embryonic stem cells identified a compound called SC1, which is capable of inhibiting both RasGAP and ERK1, and inducing differentiation both *in vitro* and *in vivo* [21]. This assay, however, was still somewhat low throughput in the context of screening since it utilized FACS analysis and was conducted using 384-well plates.

A recent screen conducted by Casalino et al. used ESC1s expressing GFP labeled βIII-tubulin to look for compounds and culture conditions that induce differentiation of ESCs into neurons [16]. A similar assay was employed using a luciferase reporter downstream of the regulatory region of neuronal Tα1 tubulin, a specific neuronal marker, upstream of the luciferase gene to identify small molecules that induce neuronal differentiation in embryonic stem cells (ESCs) [22]. This high-throughput

Table 9.1 Previously identified compounds which target DNA repair machinery

Drug/Compound Name	Target	Reference
AZD7762	CHK1	[1]
SB218078	CHK1/ATM	[2]
O^6-benzyguanine	MGMT	[6, 7]
Methoxyamine	APE1/Ref-1	[7, 9–11]
Nitroindole-2-carboxylic acid	APE1/Ref-1	[7]
Lucanthone	APE1/Ref-1	[7]
NU7026	DNA-PK	[7, 14]
IC87361	DNA-PK	[7]
IC87102	DNA-PK	[7, 15]
Wortmannin	ATM/PI3K	[7]
LY2940002	ATM/PI3K	[7]
Caffeine	ATM/PI3K	[7]

phenotypic cell-based screen of kinase-directed combinatorial libraries led to the discovery of TWS119, a 4,6-disubstituted pyrrolopyrimidine that can induce neurogenesis in murine ESCs [22]. Furthermore, the authors demonstrated that the target of TWS119 was shown to be glycogen synthase kinase-3beta (GSK-3beta), a gene previously shown to induce neurogenesis in ESCs.

In addition to screens to identify regulators of differentiation of ESCs into neurons, researchers have also begun to examine other non-neuronal differentiation endpoints. One example is the ability to differentiate human ESCs into pancreatic cells which produce insulin for the treatment of diabetes. Using a high-content chemical screen and staining for Pdx1, a gene which is significantly increased in pancreatic progenitors, a group identified a small molecule, indolactam-V, that induces differentiation of a substantial number of Pdx1-expressing cells from human ESCs [23]. The success of islet transplantation for people afflicted with type I diabetes encourages researchers to find new sources of insulin-secreting beta cells for cell replacement, including directed differentiation of ESC with compounds such as such as drugs like indolactam-V.

The use of cell imaging- based assays provides a view of the effect of compounds on the fate of stem cell differentiation at the cellular level. For example, Desbordes et al. recently developed an immunocytochemical based staining for Oct-4 in human ESCs [24]. This assay allowed the identification of molecules that promoted short-term self-renewal or drove early differentiation.

In addition to screening assays involving ESCs and differentiation, a number of groups have explored how to use differentiated stem cells to conduct screens as disease-in-a-dish models, mostly focusing on cytotoxicity assays. The most widely used cell type for these assays includes neural stem/progenitor cells and neurospheres. An advantage of working directly with neural stem (NS) cells, as discussed in a review by Danovi et al. [25], is that adherent neuronal embryonic stem (NES) cells can be grown and expanded in the presence of epidermal growth factor (EGF) and fibroblast growth factor-2 (FGF-2). Expanding ES and induced pluripotent stem (iPS) cells is a major hurdle when trying to conduct large scale screens of this nature, and NES cells bypass these technical challenges. Large scale HTS to identify compounds

Table 9.2 Summary of previously conducted HTS using stem cells

Cell type	Assay/Target	Compounds identified	Detection method	Reference
Mouse ESCs	Oct4-GFP	SC1	FACS	[21]
Mouse ESCs	βIII–tubulin	–	Fluorescence	[46]
Mouse ESCs	Neuronal Tα1 tubulin	TWS119	Luciferase	[22]
Human ESCs/ pancreatic cells	Pdx1	Indolactam-V	High content staining	[23]
Human ESCs	Oct-4 GFP	–	High content staining	[24]
Human neuronal ESCs	Proliferation	Y-27632	Luciferase	[25, 26]
Mouse SVZ stem cells	Proliferation	3 major compounds	Luciferase	[27]
iPSCs/mouse myoblasts	De-differentiation	Reversine	Live cell imaging	[28]
Glioblastoma neural stem cells	Proliferation	14 compounds	Live cell imaging	[34]
Glioblastoma neural stem cells	Proliferation	8 compounds	Luciferase	[35]
Breast cancer stem cells	Proliferation	Etoposide, salinomycin, abamectin and nigericin	Luciferase	[36]

which could enable robust proliferation of ES cultures have been conducted [25]. One of the most interesting results from this screen is the ability of Rho-associated kinase (ROCK) inhibitors to significantly expand the cultures, resulting in an increase in their self-renewal/survival. Similarly, when human ES-derived neural progenitors were transplanted *in vivo* in the presence of the ROCK inhibitor Y-27632 they demonstrated significantly increased survival [26]. In a similar HTS assay, compounds were screened to determine which, if any, have an effect on neurosphere proliferation using neuronal stem cells isolated from the adult mouse subventricullar zone (SVZ) of the hippocampus [27]. Using a relatively simple and inexpensive CellTiter-Glo based proliferation assay the authors identified compounds which not only reduce proliferation of the neurospheres, but also result in differentiation of the cells due to a programmed senescence, thus ceasing their ability to proliferate.

In addition to screening ESCs and normal (non-cancerous) tissue specific stem cells, many groups have now begun to screen induced pluripotent stem cells (iPSCs) to identify compounds which may assist in their reprogramming capabilities. This would further facilitate the use of iPS derived cells for cell therapy purposes and as relevant disease models for drug development. Skin fibroblasts or blood cells are currently reprogrammed to the pluripotent state by transfection or infection with the transcription factors Oct3/4, Sox2, Klf4 and c-Myc. To identify compounds with the equivalent effects, Chen et al. utilized an HTS assay to examine which compounds could de-differentiate mouse myoblasts [28]. The group found that the compound reversine had the ability to de-differentiate the myoblasts in pluripotent cells, and subsequently the de-differentiated cells could then be differentiated to osteoblasts and adipocytes. In a similar study, reversine could also induce pluripotency in mouse and human fibroblasts such that the cells could then be differentiated into skeletal muscle [29].

Potential drug candidates are routinely tested for toxic effects in liver-derived hepatocytes and heart derived cardiomyocytes as pre-clinical models of toxicity (reviewed in [30]). The use of differentiated cells for these assays is not ideal and although primary cells would provide a closer cellular model to testing *in vivo*, they are not readily amenable for routine use in larger scale testing. To circumvent these limitations, both hepatocytes and cardiomyocytes have been derived from human iPSCs and are being investigated in small scale drug screens as more relevant cellular models [31, 32]). However, the cellular heterogeneity and low differentiation efficiency of the iPS cells [33] still presents a huge hurdle for these types of screens.

The ability to screen ESCs, normal tissue specific stem cells and more recently iPSCs is providing the foundation to develop HTS assays to screen with cancer stem cells or CSCs. To date only a handful of screens have been conducted using CSCs. One of the first screens utilized CSCs isolated from the brain, specifically tumor cells from patients with glioblastoma (GNS: glioblastoma neural stem cells) [34] to identify compounds which blocked self-renewal, either by inducing cell death or differentiation. The authors used a live-imaging-based assay to screen a collection of 450 compounds already in clinical use (NIH clinical collection) against 3 primary GNS lines, followed up with an additional screen using 1,000 chemicals with one of the GNS cell lines. Form the later screen, a total of 14 compounds had the desired activity, including kinase inhibitors against platelet-derived growth factor (PDGF) and insulin-like growth factor-1 (IGF-1) receptor tyrosine kinases (RTK). In a similar screen, patient derived GNS cells were screened against a 30,000 small molecule library [35] to find cytotoxic compounds. The cells were cultured in a monolayer, yet were grown in a stem cell based sphere media to enrich for CSCs, and cytotoxic compounds were detected using ATP-based cell viability assay. A total of 694 hits were indentified and further evaluated using secondary assays. Eight compounds were found to preferentially inhibit the GNS cells compared to the differentiated counterparts. Mechanistically, these active compounds down-regulated GNS associated genes, but further validation is required to characterize their mechanism of action.

The most significant study to date using CSCs and HTS technology was conducted in a breast cancer model by Gupta et al. in 2009 [36]. Due to the difficultly of growing CSCs for large scale screens this group employed a model of epithelial-mesenchymal transition (EMT) to generate their cells. The process of EMT has been found to activate the same transcription factors that provide cancer cells with the ability to self-renew, resist apoptosis, generate tumorspheres and induce tumors at low numbers compared to control cells when injected into mice. In addition, cells which have undergone an EMT have been shown to be the most aggressive cells in a tumor and have been further characterized as one population of CSCs [37–39]. To generate these cells the group utilized normal human mammary epithelial cells (HMECs) which have been transformed to a tumorigenic state via the introduction of SV40 large-T antigen, the telomerase catalytic subunit hTERT and H-Ras to generate HMLER breast cells [40]. The HMLER cells were then infected with a short hairpin RNA against E-cadherin, a gene known to regulate the epithelial state of cells and when knocked down an aggressive mesenchymal phenotype is produced [41].

The loss of E-cadherin led to an increase in the expression of the CSC maker profile CD44^{high}/CD24^{low}, mammopshere formation and an increased resistance to two commonly used chemotherapeutic drugs paclitaxel and doxorubicin. The HMLER^{shEcad} and HMLER^{shCntrl} lines were both screened in a 384 well microplate format and assayed for viability using CellTiter-Glo. A total of 16,000 compounds were tested and only 32 compounds (0.2 % of the total library assayed) resulted in selective toxicity toward the HMLER^{shEcad} cells and not the HMLER^{shCntrl} line. Upon re-testing, only 8 of the 32 compounds demonstrated reproducible toxicity and selectivity. The list of eight compounds included etoposide, salinomycin, abamectin and nigericin. Further characterization of the antibiotic salinomycin was carried out because treatment with the drug led to a decrease in the CD44^{high}/CD24^{low} population within the HMLER^{shEcad} cells. Treatment of HMLER^{shEcad} cells with salinomycin also resulted in a decrease in tumorsphere formation and although treatment of the non-tumorigenic HML-Mx cells decreased mammosphere formation, it did not affect the number relative to control. Furthermore, cell proliferation with salinomycin was not inhibited when treating cells grown in monolayer indicating that this effect was not simply a result in the inhibition of cell proliferation. Additional assays in the SUM159 and 4T1 breast cancer models were conducted demonstrating the effect of salinomycin at inhibiting CSC cell growth and metastasis *in vivo*. This study remains one of the best examples of screens to find compounds that selectively target CSC viability and sets the foundation for the development of additional assays.

9.4 Access to HTS Capabilities for Large Scale Screening and Drug Development

Building a high throughput screening facility requires a significant investment in infrastructure, including equipment, compound collections, a laboratory informatics management system (LIMS), and automation and assay development expertise, and in most cases access to medicinal chemistry support to improve the potency and selectivity of the original actives found from the HTS. Access to HTS capabilities by the broader scientific community is therefore not trivial. The Society of Laboratory Automation and Screening provides a list of academic and government (NCATS) HTS laboratories. Most of these HTS laboratories screen libraries of chemically synthesized compounds (in contrast to extracted natural products) that are commercially available. The largest drug collection publically available is from the NIH Molecular Libraries Small Molecule Repository (MLSMR) with ~350,000 compounds. In addition to the MLSMR libraries, there are many smaller non-MLSMR collections being acquired from commercial and academic resources to increase the diversity and to better address so-called undruggable targets. These compound libraries are better suited to screen difficult assays such as those involving stem cells. These libraries include the Library of Pharmacologically Active Compounds (LOPAC) from Sigma, Prestwick libraries, National Cancer Institute (NCI) diversity set and more recently the in-house NCATS (NCGC) pharmaceutical collection, also called the NPC [42].

The NPC represents a complete and non-redundant collection of all molecular entities approved for clinical use in the United States, England, Canada, Europe and Japan. In addition, compounds that have been registered for human testing, but not 100 % approved by a regulatory agency were also included. This list included drugs registered by the United States Drug Enforcement Agency, compounds listed in the World Health Organization International Nonproprietary Name and the United States Adopted Names registries, and finally, compounds listed as drugs that have been approved by a New Drug (IND) application. Screening these collections enable the possibility of finding new indications for existing drugs (so-called drug repurposing), which potentially can streamline the approval process for testing in the clinic. NIH provides funding specifically to help develop assays for HTS (RO1), implementation of screens (RO1) and medicinal chemistry improvement of already existing compounds. To assist in the translation from screening and basic research to the bedside, the NIH has also established several programs, including, Therapeutics for Rare and Neglected Diseases (TRND), NexT, Blueprint, and Rapid Access to Interventional Development (RAID) that are critical resources needed for the pre-clinical development of new therapeutic agents.

In addition to the NCATS pharmaceutical collection, additional custom small molecule collections are being assembled that target oncology-relevant pathways/mechanisms of survival. These compounds are in various levels of pre-clinical and clinical testing, and although the path to clinic might not be as direct, the biological annotation of the compounds can provide information on critical pathways regulating CSCs survival. We are currently screening such collections using a CSC cell growth assay in 1536-well microplate format (in press, *Journal of Biomolecular Screening*). The CSCs were generated using the spheroid technique since it affords for the production of large amounts of CSCs. The assay was developed using both pancreatic and prostate cancer CSCs, and overall the data demonstrated a decent number of hits which potency in the nanomolar range. This assay, however, only measures cell growth and toxicity and does not actually measure DNA repair mechanisms.

In most cases, DNA repair pathways assays are carried out using immunocytochemical staining for foci accumulation of repair proteins such as H2AX and Rad51. These assays can be performed directly in multi-well plates using standard fixation and staining protocols. The images are quantified using fluorescent microscopy and image analysis software. This so-called high content screening provides pathway specific information on a per cell basis, and assays can be implemented in 96-, 384- and 1536-well microplate format. However, the automation protocols are complex because of the high number of addition and washes during the fixation and staining steps. For this reason, in practice, high content screens are in most cases applied to <100,000 compound libraries. Applying these already developed DNA repair high content imaging assays such as the H2AX and Rad51 foci accumulation to CSCs should provide the next generation assays needed to exploit these pathways.

Another widely used method to access DNA damage utilizes a GFP mediated assay to quantify interstrand cross-linked induced HR in mammalian cells based on the DR-GFP reporter. This reporter detects HR induced by DNA double-strand breaks

(DSBs) [43]. DR-GFP is composed of two differentially mutated green fluorescent protein (GFP) genes oriented as direct repeats, which is where the DR notation is derived from. The upstream repeat contains the recognition site for the rare-cutting I-*Sce*I endonuclease and the downstream repeat is a 5′ and 3′ truncated GFP fragment. Upon transient expression of I-*Sce*I this leads to a DSB in the upstream GFP gene and HR to repair the DSB resulting in GFP^+ cells which are quantified by flow cytometry. To date, this assay has been widely used to identify proteins required for HR repair, such as BRCA1 and BRCA2, and to determine which pathways suppress HR repair, using both candidate gene approaches and whole genome screens [44]. Conducting an assay such as this in HTS format would require an HTS compatible fluorescent cytometry (HTFC). CSCs infected with the DR-GFP plasmid could be challenged with compounds and then assayed using the HTFC for their ability to repair the breaks. This could prove to be an efficient tool for the rapid screening of CSC DNA repair when HT compatible flow cytometer instruments are available. In this regard, new instruments such as the IntelliCyt's new HTFC screening system might be able to enable 384-well microplate screening using this type of assays, perhaps even screening of libraries of more than 10,000 compounds.

Other non-HTS friendly assays are also used to measure DNA repair pathways in numerous cell types. The most frequently used is the Comet assay which measures DNA strand breaks in cells which have been embedded in an agarose gel. The embedded cells are lysed with detergent and high salt to form nucleoids containing supercoiled loops of DNA linked to the nuclear matrix. When the gel is electrophoresed at high pH it results in structures resembling comets which are observed and quantified by fluorescence microscopy. These images are quantified based on the intensity of the comet tail relative to the head and this reflects the number of DNA breaks. If the nucleoids are incubated with bacterial repair endonucleases that recognize specific kinds of damage in the DNA and convert lesions to DNA breaks this increases the sensitivity and specificity of the assay, and further increases the amount of DNA in the comet tail. This assay has been employed to study how effective CSCs isolated from the pancreas repair DNA after being treated with gemcitabine compared to bulk tumor cells [45]. The established lines HPAC and PANC1 were compared to bulk cells, and it was found that CSCs isolated from these lines functionally repair breaks in DNA faster after challenged with the drug gemcitabine [45]. Ideally we would like to use this assay to conduct HTS for compounds that inhibit DNA repair. However, presently, because of the extraction and separation steps, this comet assay is a very low throughout and would have to be reformatted to be able to use for HTS. However, it can be used as a secondary assay to validate the hits obtained from higher throughout assays such us those described above.

9.5 Conclusions

The development of protocols to produce and expand CSCs is enabling their use in HTS settings to find small molecules that affect CSC viability. Using the spheroid technique of culturing CSCs and a battery of HTS-compatible assays ranging from cell growth to probing DNA repair pathways by high content imaging, we aspire to find compounds which can eradicate these highly aggressive cells from bulk populations of tumors. We speculate that the compounds can be identified that will target the efficient DNA repair machinery of these cells. Many of the compound libraries previously mentioned contain checkpoint kinase inhibitors and future studies examining drugs which synergize with these inhibitors are underway. This research will be first conducted using models of pancreatic cancer because of the limited treatment options available for this highly aggressive disease, but we hope to expand the same HTS and drug development paradigm to other forms of cancer.

References

1. Frosina G (2009) DNA repair in normal and cancer stem cells, with special reference to the central nervous system. Curr Med Chem 16(7):854–866
2. Gallmeier E, Hermann PC, Mueller MT et al (2011) Inhibition of ataxia telangiectasia- and Rad3-related function abrogates the in vitro and in vivo tumorigenicity of human colon cancer cells through depletion of the CD133(+) tumor-initiating cell fraction. Stem Cells 29(3):418–429
3. Garrett MD, Collins I (2011) Anticancer therapy with checkpoint inhibitors: what, where and when? Trends Pharmacol Sci 32(5):308–316
4. Wolpaw AJ, Shimada K, Skouta R et al (2011) Modulatory profiling identifies mechanisms of small molecule-induced cell death. Proc Natl Acad Sci USA 108(39):E771–E780
5. Lehar J, Krueger AS, Avery W et al (2009) Synergistic drug combinations tend to improve therapeutically relevant selectivity. Nat Biotechnol 27(7):659–666
6. Ranson M, Middleton MR, Bridgewater J et al (2006) Lomeguatrib, a potent inhibitor of O6-alkylguanine-DNA-alkyltransferase: phase I safety, pharmacodynamic, and pharmacokinetic trial and evaluation in combination with temozolomide in patients with advanced solid tumors. Clin Cancer Res 12(5):1577–1584
7. Kelley MR, Fishel ML (2008) DNA repair proteins as molecular targets for cancer therapeutics. Anticancer Agents Med Chem 8(4):417–425
8. Kato T, Natsume A, Toda H et al (2010) Efficient delivery of liposome-mediated MGMT-siRNA reinforces the cytotoxity of temozolomide in GBM-initiating cells. Gene Ther 17(11):1363–1371
9. Fishel ML, He Y, Smith ML et al (2007) Manipulation of base excision repair to sensitize ovarian cancer cells to alkylating agent temozolomide. Clin Cancer Res 13(1):260–267
10. Rinne M, Caldwell D, Kelley MR (2004) Transient adenoviral N-methylpurine DNA glycosylase overexpression imparts chemotherapeutic sensitivity to human breast cancer cells. Mol Cancer Ther 3(8):955–967
11. Taverna P, Liu L, Hwang HS et al (2001) Methoxyamine potentiates DNA single strand breaks and double strand breaks induced by temozolomide in colon cancer cells. Mutat Res 485(4):269–281
12. Sengupta S, Harris CC (2005) p53: traffic cop at the crossroads of DNA repair and recombination. Nat Rev Mol Cell Biol 6(1):44–55
13. Zhuang W, Li B, Long L et al (2011) Knockdown of the DNA-dependent protein kinase catalytic subunit radiosensitizes glioma-initiating cells by inducing autophagy. Brain Res 1371:7–15
14. Nutley BP, Smith NF, Hayes A et al (2005) Preclinical pharmacokinetics and metabolism of a novel prototype DNA-PK inhibitor NU7026. Br J Cancer 93(9):1011–1018
15. Shinohara ET, Geng L, Tan J et al (2005) DNA-dependent protein kinase is a molecular target for the development of noncytotoxic radiation-sensitizing drugs. Cancer Res 65(12):4987–4992

16. Tichy ED, Stambrook PJ (2008) DNA repair in murine embryonic stem cells and differentiated cells. Exp Cell Res 314(9):1929–1936
17. Ralhan R, Kaur J, Kreienberg R et al (2007) Links between DNA double strand break repair and breast cancer: accumulating evidence from both familial and nonfamilial cases. Cancer Lett 248(1):1–17
18. Inglese J, Auld DS, Jadhav A et al (2006) Quantitative high-throughput screening: a titration-based approach that efficiently identifies biological activities in large chemical libraries. Proc Natl Acad Sci USA 103(31):11473–11478
19. Collins FS (2011) Reengineering translational science: the time is right. Sci Transl Med 3(90):1–6
20. Severyn B, Liehr RA, Wolicki A et al (2012) Parsimonious discovery of synergistic drug combinations. ACS Chem Biol 6(12):1391–1398
21. Chen S, Do JT, Zhang Q et al (2006) Self-renewal of embryonic stem cells by a small molecule. Proc Natl Acad Sci USA 103(46):17266–17271
22. Ding S, Wu TY, Brinker A et al (2003) Synthetic small molecules that control stem cell fate. Proc Natl Acad Sci USA 100(13):7632–7637
23. Chen S, Borowiak M, Fox JL et al (2009) A small molecule that directs differentiation of human ESCs into the pancreatic lineage. Nat Chem Biol 5(4):258–265
24. Desbordes SC, Placantonakis DG, Ciro A et al (2008) High-throughput screening assay for the identification of compounds regulating self-renewal and differentiation in human embryonic stem cells. Cell Stem Cell 2(6):602–612
25. Danovi D, Falk A, Humphreys P et al (2010) Imaging-based chemical screens using normal and glioma-derived neural stem cells. Biochem Soc Trans 38(4):1067–1071
26. Koyanagi M, Takahashi J, Arakawa Y et al (2008) Inhibition of the Rho/ROCK pathway reduces apoptosis during transplantation of embryonic stem cell-derived neural precursors. J Neurosci Res 86(2):270–280
27. Liu Y, Lacson R, Cassaday J et al (2009) Identification of small-molecule modulators of mouse SVZ progenitor cell proliferation and differentiation through high-throughput screening. J Biomol Screen 14(4):319–329
28. Chen S, Zhang Q, Wu X et al (2004) Dedifferentiation of lineage-committed cells by a small molecule. J Am Chem Soc 126(2):410–411
29. Anastasia L, Sampaolesi M, Papini N et al (2006) Reversine-treated fibroblasts acquire myogenic competence in vitro and in regenerating skeletal muscle. Cell Death Differ 13(12):2042–2051
30. Phillips BW, Crook JM (2010) Pluripotent human stem cells: a novel tool in drug discovery. BioDrugs 24(2):99–108
31. Baxter MA, Rowe C, Alder J et al (2010) Generating hepatic cell lineages from pluripotent stem cells for drug toxicity screening. Stem Cell Res 5(1):4–22
32. Inoue H, Yamanaka S (2011) The use of induced pluripotent stem cells in drug development. Clin Pharmacol Ther 89(5):655–661
33. Kim K, Doi A, Wen B et al (2010) Epigenetic memory in induced pluripotent stem cells. Nature 467(7313):285–290
34. Pollard SM, Yoshikawa K, Clarke ID et al (2009) Glioma stem cell lines expanded in adherent culture have tumor-specific phenotypes and are suitable for chemical and genetic screens. Cell Stem Cell 4(6):568–580
35. Visnyei K, Onodera H, Damoiseaux R et al (2011) A molecular screening approach to identify and characterize inhibitors of glioblastoma multiforme stem cells. Mol Cancer Ther 10(10):1818–1828
36. Gupta PB, Onder TT, Jiang G et al (2009) Identification of selective inhibitors of cancer stem cells by high-throughput screening. Cell 138(4):645–659
37. Iwatsuki M, Mimori K, Yokobori T et al (2010) Epithelial-mesenchymal transition in cancer development and its clinical significance. Cancer Sci 101(2):293–299
38. Kong D, Banerjee S, Ahmad A et al (2010) Epithelial to mesenchymal transition is mechanistically linked with stem cell signatures in prostate cancer cells. PLoS One 5(8):e12445

39. Singh A, Settleman J (2010) EMT, cancer stem cells and drug resistance: an emerging axis of evil in the war on cancer. Oncogene 29(34):4741–4751
40. Elenbaas B, Spirio L, Koerner F et al (2001) Human breast cancer cells generated by oncogenic transformation of primary mammary epithelial cells. Genes Dev 15(1):50–65
41. Onder TT, Gupta PB, Mani SA et al (2008) Loss of E-cadherin promotes metastasis via multiple downstream transcriptional pathways. Cancer Res 68(10):3645–3654
42. Huang, R, Southall N, Wang Y et al (2011) The NCGC pharmaceutical collection: a comprehensive resource of clinically approved drugs enabling repurposing and chemical genomics. Sci Transl Med 3(80):80ps16
43. Akyuz N, Boehden GS, Susse S et al (2002) DNA substrate dependence of p53-mediated regulation of double-strand break repair. Mol Cell Biol 22(17):6306–6317
44. Nakanishi K, Cavallo F, Brunet E et al (2004) Homologous recombination assay for interstrand cross-link repair. Methods Mol Biol 745:283–291
45. Mathews LA, Cabarcas SM, Hurt EM et al (2011) Increased expression of DNA repair genes in invasive human pancreatic cancer cells. Pancreas 40(5):730–739
46. Casalino L, Magnani D, De Falco S et al (2012) An automated high throughput screening-compatible assay to identify regulators of stem cell neural differentiation. Mol Biotechnol 50(3):171–180

Chapter 10
The Future of DNA Repair and Cancer Stem Cells

Stephanie M. Cabarcas

Abstract The existence of cancer stem cells (CSCs), tumor-initiating cells (TICs) or cancer-initiating cells (CICs) has been a topic of hot debate in the field of cancer biology. The molecular mechanism(s) which this specific subpopulation of cells utilize to sustain themselves is currently under investigation. Recent studies demonstrate that CSCs express enhanced DNA repair mechanisms which can contribute to their ability to evade traditional cancer treatments. The following is a summary of the previous chapters presented in this text with an emphasis on the future of CSCs and their potential contribution to the development and identification of novel therapies and targets.

Cancer development is characterized by uncontrolled cellular growth that is derived from genetic mutations that alter normal homeostatic regulation. The ability of cancer cells to drive tumor formation and create a mass that is comprised of a variety of cells with distinct genetic and epigenetic gene signatures demonstrates tumor heterogeneity. As described in intimate detail throughout this text, the existence of a subset of tumor cells termed 'cancer stem cells' (CSCs), 'tumor-initiating cells' (TICs) or 'cancer-initiating cells' (CICs), has given the field of cancer biology a new avenue and inroad to investigate additional mechanism(s) by which cancer can be targeted. The hierarchy model of cancer hypothesizes that there are biologically distinct cells within a tumor that possess the capability to drive tumor formation, metastasis and provide a mechanism for resistance during treatment [1]. The ability to resist and survive treatments ultimately results in patient downfall. It has been repeatedly demonstrated that this population and these deadly traits can be attributed to the CSC [1–3]. A detailed introduction to CSCs and their role in cancer biology is presented, as well in combination with the mechanism(s) and pathways that have evolved in the CSC to work in evasion of traditional cancer treatments including radiation and chemotherapies.

The fundamental regulatory pathways which enable the CSC to escape targeting by these 'attacks' is under intense investigation. Recently, laboratories have observed that the CSCs display a key trait which enhances their ability to escape assault and this is the increase in their genomic stability (reviewed in

S. M. Cabarcas (✉)
Department of Biology, Gannon University, Erie, Pennsylvania, USA
e-mail: cabarcas002@gannon.edu

L. A. Mathews et al. (eds.), *DNA Repair of Cancer Stem Cells*,
DOI 10.1007/978-94-007-4590-2_10,

[4]). It is this increase in genomic stability and capacity to withstand insult by various forms of cancer therapies that contribute to their maintenance in comparison to the bulk population of cells which are often successfully targeted. The ability of CSCs to display an increased genomic stability is coupled with the ability to display an increased and more efficient means to utilize DNA repair mechanisms.

The DNA repair mechanisms utilized by cells have developed to ensure the maintenance of essential cellular processes which can be disrupted by genetic insult and damaging agents. These pathways are discussed in great detail in various chapters in this text in combination with the molecular pathways which function in their repair. These specific mechanisms include base-excision repair (BER), mismatch repair (MMR), nucleotide excision repair (NER), homologous recombination (HR) and non-homologous end joining (NHEJ). The ability to utilize various repair mechanisms, which are specific to the type of DNA damage that is incurred, is a necessary homeostatic process which the CSC has manipulated to ensure survival. The maintenance of genomic integrity is essential to proper development and molecular pathways which govern necessary cellular processes. An example of how these pathways are utilized to ensure proper development is eloquently described in Chap. 4 (written by Olga Momčilović and Gerald Schatten), in relation to their role in normal stem cells. Momčilović and Schatten describe in detail how stem cells, including pluripotent, embryonic and multipotent stem cells, evade damage by utilizing various DNA repair pathways such as the double-stranded break repair pathways (DSB): HR and NHEJ; and NER, BER, MMR mechanisms as well. The use of these mechanisms to protect and avoid acquisition of DNA mutations that can be passed from generation to generation is essential for survival. This in-depth and detailed account of how stem cells utilize these pathways to preserve their genome is critical to understanding how CSCs may function. In addition to focusing on DNA repair pathways, various chapters discuss the role of quiescence and its importance in minimizing cellular oxidative stress by influencing the type of DNA repair pathways which is utilized, favoring NHEJ, an error prone pathway. Hence, the concept of quiescence in aiding genomic integrity is considered a double-edged sword but, it is apparent that an insufficient DNA repair system in stem cells contributes to gain of mutation that result in disease and malignancy. Thus, these processes much be strictly regulated and remain free of error.

The need to further elucidate the molecular mechanism(s) by which CSCs sustain themselves in a lethal and harsh environment created by anti-cancer therapies is critical to furthering the clinical aspect of CSC research. The long-term goal to utilize quantitative high-throughput (qHTS) methods to identify potential targets and drugs which can result in CSC differentiation or death has potential to be extremely fruitful and significant. As demonstrated in this text, there are various drugs readily available which have the potential to target CSCs and result in successful patient treatment in regards to metastasis and chemo- and radio-resistant cancers. These drugs vary from naturally derived anti-cancer compounds to drugs which are capable of targeting DNA repair pathways. In the case of CSCs, the enhanced DNA repair pathways is a critical contributor to their survival, thus, it is beneficial to screen drugs capable of shutting these pathways down. It is quite promising that these drugs targeting DNA repair pathways administered as adjuvant therapy with traditional anti-cancer therapies can result in increased survival and eradication of resistant CSCs.

There is evidence, albeit modest, that a major characteristic of the CSC that allows for survival is its ability to utilize DNA repair pathways and resistance mechanisms. The cancers which this hypothesis has been tested include leukemia, colon, lung, glioblastoma, breast and pancreatic. The CSCs isolated from these models exhibit enhanced resistance pathways and DNA repair mechanisms. This text successfully compiles all lines of evidence and data which support and call for further investigation into this specific aspect of CSC regulation. In addition to providing an in-depth discussion of the various pathways which regulate this survival mechanism of CSCs, we focus on means of targeting this population via interruption of enhanced DNA repair pathways as well. The common thread which connects these chapters is not only the focus on the exact pathways which CSCs utilize, but the call for development or use of treatments which target this aspect of the CSC. As established, the CSC is a driving force of tumor initiation and metastasis. The ultimate demise of patients suffering from aggressive and deadly cancers is a consequence of metastasis resulting from resistance to traditional treatment. In a time where the major focus of the cancer biology community is to propel research into a place where we can translate bench to bedside, the need to further investigate CSCs and its underlying molecular biology comes to play. The existence of CSCs has been a topic of heated debate for the last few years but with overwhelming evidence and data, it is now apparent that this population exists and is a major contributor to cancer development, metastasis and resistance. The evidence which demonstrates CSCs have an increase in their genomic stability and results in survival under stressful and attacking conditions is presented throughout this text and is exemplified in various cancer models. The concepts discussed in this text demonstrate that this aspect of the CSC field deserves further attention and can provide an additional path and perspective in the field of molecular cancer therapeutics and development of treatment. We believe this field should gain momentum and can greatly impact the efforts being put forth to work towards eradicating these highly aggressive and deadly cancers.

References

1. Lobo NA et al (2007) The biology of cancer stem cells. Annu Rev Cell Dev Biol 23;675–699
2. Hurt EM, Farrar WL (2008) Cancer stem cells: the seeds of metastasis? Mol Interv 8(3):140–142
3. Tang C, Ang BT, Pervaiz S (2007) Cancer stem cell: target for anti-cancer therapy. FASEB J 21(14):3777–3785
4. Mathews L, Cabarcas S, Farrar W (2011) DNA repair: the culprit for tumor-initiating cell survival? Cancer Metast Rev 30(2):185–197

Index

L. A. Mathews et al. (eds.), *DNA Repair of Cancer Stem Cells*,
DOI 10.1007/978-94-007-4590-2, © Springer Science+Business Media Dordrecht 2013

R

S

T

V

W

X

Z

Printed by Books on Demand, Germany